About The Author

Presh Talwalkar studied Economics and Matl
University. His site *Mind Your Decisions* has [illegible] posts and original videos about math that have been viewed millions of times.

Books By Presh Talwalkar

The Joy of Game Theory: An Introduction to Strategic Thinking. Game Theory is the study of interactive decision-making, situations where the choice of each person influences the outcome for the group. This book is an innovative approach to game theory that explains strategic games and shows how you can make better decisions by changing the game.

Math Puzzles Volume 1: Classic Riddles in Counting, Geometry, Probability, and Game Theory. This book contains 70 interesting brain-teasers.

Math Puzzles Volume 2: What do an Infinite Tower, a Classic Physics Puzzle, and Coin Flipping Have in Common? This is a follow-up puzzle book with 45 delightful puzzles.

But I only got the soup! This fun book discusses the mathematics of splitting the bill fairly.

40 Paradoxes in Logic, Probability, and Game Theory. Is it ever logically correct to ask "May I disturb you?" How can a football team be ranked 6th or worse in several polls, but end up as 5th overall when the polls are averaged? These are a few of the thought-provoking paradoxes covered in the book.

Multiply By Lines. It is possible to multiply large numbers simply by drawing lines and counting intersections. Some people call it "how the Japanese multiply" or "Chinese stick multiplication." This book is a reference guide for how to do the method and why it works.

Table of Contents

Why Learn Mental Math Tricks?

Mental math has a mixed reputation. Some consider it useless because calculators and computers can solve problems faster, with assured accuracy. Additionally, mental math is not even necessary to get good grades in math or to pursue a professional math career. So what's the point of learning mental math and math tricks anyway?

There are many reasons why mental math is still useful. For one, math skills are needed for regular tasks like calculating the tip in a restaurant or comparison shopping to find the best deal. Second, mental math tricks are one of the few times people enjoy talking about math. Third, mental math methods can help students build confidence with math and numbers.

Mental math tricks are fun to share. Imagine your friend asks you to multiply 93 and 97, and before they can type it into a calculator, you can figure out the answer in your head. Or imagine your friend asks you to divide 17 by 91, and you can calculate out the exact decimal answer in your head. Mental math tricks are impressive to see and they can be exciting to learn.

Mental math reduce the chore of simple math, build confidence in math skills, and serve as educational entertainment at parties and pubs. The flip side is some mental math calculations require extensive memorization and lots of practice to learn. This book is therefore not a comprehensive collection of mental math methods. This book seeks the middle ground: tricks that are relatively easy to learn, are fun, or have educational value.

Each section explains a math trick, and then has practice problems with complete solutions to illustrate the method. Every method is accompanied with a mathematical proof to justify why the trick works. While I find the proofs the most interesting parts, they can be skipped without detriment to learning the methods. The book can be read from start to finish. Or you can skip around, as most sections are independent of each other, excluding a few tricks that build upon previous ones.

Part I: Introductory Tricks

Calculate Percentages

While you can use a calculator to figure out percentages, there is a simple method you can use to solve them in your head. The best part is you probably are familiar with the basic skills for calculating percentages, which are calculating 10% and 50%.

Calculating 10%. Calculating 10 percent of a number is the same as dividing the number by 10. The procedure to divide a number by 10 is to move the decimal point one spot to the left. For example, to calculate 10% of 123 is the same as dividing 123 by 10. This is accomplished by moving the decimal point over 1 spot to the left, which means that 12.3 is 10% of 123. Similarly 10% of 56 is 5.6 and 10% of 4,987.56 is 498.756.

Calculating 50%. Calculating 50 percent of a number is the same as dividing the number by 2, which means to halve the number. For example, to calculate 50% of 123 is the same as halving 123. So 50% of 123 is 61.5. Similarly 50% of 56 is 28 and 50% of 5,000 is 2,500.

Calculating Other Percentages. You can calculate most percentages in your head if you can calculate 10% and 50%! The reason is that many percentages can be found as a combination of 10% and 50%.

Let's do an example of calculating 15% of 63. Note that 15% is equal to 10% plus 5%. To do this, you can calculate 10% and then you can add half of 10% (which is 5%) to itself. So we calculate 10% of 63 is equal to 6.3. Then we calculate 5% as half of that, which is 3.15. So we add these together to get 9.45, which is 15% of 63.

Let's do another example of 25% of 84. Note that 25% is equal to half of 50%. So we can calculate 25% of 84 by finding 50% and then halving the result. We can readily calculate that 50% of 84 is equal to 42, and then we need to take half of that number to get 21. Therefore we have 25% of 84 is equal to 21.

It is easy to calculate 10% and 50% in your head. From those starting points, you can easily find 5% (which is half of 10%), and then you can calculate all of the common percentages that are multiples of 5, 10, or 25.

Here is a short guide.

5% = half of 10%

10% = divide by 10

15% = add 10% and half of 10%

20% = find 10% and then double it

25% = find 50% and then halve it

30% = find 10% and then triple it

40% = find 10% and 50%. Subtract 10% value from 50% value

50% = divide in half

60% = find 10% and 50%. Add 10% value to 50% value

75% = find 50% and 25% and add together

80% = find 20% (as double 10%) and subtract from the whole

90% = find 10% and subtract from the whole

There can be more than one way to calculate a given percentage. For example, you can also calculate 60% by first finding 10% and then multiplying that value by 6.

The point is that you can calculate many common percentages by knowing how to calculate 10% and 50% in your head—and you probably can do these things without too much difficulty.

This trick can also help you estimate. For example, let's say you want to find 35% of 4,987. We can round 4,987 to a value of 5,000. Now we can calculate 10% is equal to 500, and 5% is half of that, or 250. So we can calculate 30% is triple 10% and equal to 1,500. We then add 5% and get to 1,750. This is very close to the exact answer of 1,745.45 for 35% of 4,987.

Now let's solve some practice problems.

Practice Problems

25% of 88

15% of 65

75% of 92

30% of 168

90% of 65

Proof

The method works because we can decompose a percentage like 15% into 15% = 10% + 5% = 10% + (10%)/2. In other words, it works because we can decompose many percentages into the operations of calculating 10% and calculating 50%.

Solutions to Practice Problems

25% of 88. We first calculate 50% by halving 88 to get 44. Then we divide that in half to find 25% is equal to 22.

15% of 65. We first calculate 10% which is 6.5, and then we divide that in half to find 5% is equal to 3.25. We add those results to get 15% of 65 is equal to 9.75.

75% of 92. We first calculate 50% by halving 92 to get 46. Then we divide that in half to find 25% is equal to 23. Subtracting 25% from the whole gets us that 75% of 92 is equal to 69.

30% of 168. We first calculate 10% which is 16.8, and we halve to get 5% is 8.4. Since 50% of 168 is 84, that means 25% is half of that, or 42. We add the 25% and 5% to get that 30% of 168 is 50.4.

90% of 65. We first calculate 10% by moving the decimal point to get 6.5. Then we subtract 10% from the whole to get 90% of 65 is equal to 58.5.

The Rule Of 72

You invest \$1,000 at a guaranteed 4% interest rate annually. How long will it take your investment to double to \$2,000?

We can solve the problem by computing compound interest. In the first year, you earn 4% on \$1,000 which is \$40. In other words, your investment grows to \$1,000(1.04) = \$1,040. In the next year, you earn another 4% on top of \$1,040, which means your investment grows to \$1,040(1.04) = \$1,081.6 = $\$1{,}000(1.04)^2$. Noting the pattern, we can see that in T years your investment will grow to $\$1{,}000(1.04)^T$. We can set this equation equal to \$2,000 and solve (using logarithms) that T is about 17.67 years. In other words, it takes about 18 years for your investment to double.

The Rule of 72 is a shortcut to estimate this answer very quickly. You divide 72 by the interest rate, expressed as a percentage, to get the doubling time.

For example, to calculate the doubling time with a fixed 4% interest rate, we divide 72 by the interest rate of 4 percent. Since 72/4 = 18, that means the investment will double in roughly 18 years. This is pretty close to the exact answer of 17.67!

Similarly, we can solve that an investment with 6% interest will double in roughly 72/6 = 12 years, an investment with 8% interest will double in roughly 72/8 = 9 years, and an investment with 12% interest will double in roughly 72/12 = 6 years. These are all very close to the exact answers, and the rule avoids having to compute compound interest.

There are two important caveats when using the rule of 72. First, the rule is a good approximation but it is not an exact answer. When you are working with investments, you may want to check an exact answer with a calculator or spreadsheet.

Second, the rule of 72 works only if the interest rate is fixed. When you invest in the stock market, the rate of return varies over time, and so you cannot use the rule of 72 with the same accuracy as you could for an investment with a fixed and guaranteed return.

Practice Problems

Calculate the doubling times for the following fixed interest rates.

2%

5%

7%

10%

20%

Proof

The rule works because it is an approximation to the compound interest formula. In order for an investment of *R* percent to double in *T* years, we need the interest multiplier $(1 + R)^T$ to be equal to 2. We can solve for *T* by using the natural logarithm to get $T = \ln(2)/\ln(1 + R)$. We then use the approximations that $\ln(2) = 0.69$ and $\ln(1 + R) = R$ (for values of *R* less than 30 percent). Therefore, we have $T = 0.69/R$.

Now we multiply by 100/100 to get rid of the decimal points. Thus, we have $T = 69/(100R)$, so we can consider the interest rate *R* as a whole number instead of a percentage (4% is just 4). Finally we get a slightly better approximation, and have more factors to divide by evenly, when we make the numerator 72 instead of 69. So we have $T = 72/(100R)$.

Solutions to Practice Problems

2% will double in about 72/2 = 36 years

5% will double in about 72/5 = 14.4 years

7% will double in about 72/7 = 10.3 years

10% will double in about 72/10 = 7.2 years

20% will double in about 72/20 = 3.6 years

Estimate Your Hourly Wage

Let's say you earn $50,000 in a full-time job. What is your hourly wage? A good estimate is to divide your compensation by 2,000. This can be done expediently by halving your compensation and then dividing by 1,000 (moving the decimal 3 spots to the left).

So let's divide $50,000 by 2,000. We halve $50,000 to get $25,000, and then we divide by 1,000 to get $25 per hour.

Practice Problems

Calculate the hourly wage for the following annual incomes.

$20,000

$60,000

$100,000

$250,000

$1,000,000

Proof

In America, a full-time job typically means 40 hours per week for about 50 weeks, which translates into 40 x 50 = 2,000 hours in a year. So you can find your hourly wage by dividing your annual salary by 2,000.

Solutions to Practice Problems

$20,000 is about an hourly wage of $20,000/2,000 = $10

$60,000 is about an hourly wage of $60,000/2,000 = $30

$100,000 is about an hourly wage of $100,000/2,000 = $50

$250,000 is about an hourly wage of $250,000/2,000 = $125

$1,000,000 is about an hourly wage of $1,000,000/2,000 = $500

Calculate The Day Of The Week For A Date

January 1, 2015 fell on a Thursday. Which day of the week corresponds to the date of December 25, 2015?

There are many ways to find the answer. Perhaps the simplest is to look up the answer in a calendar, a spreadsheet, or a website like WolframAlpha. But it is actually not too difficult to figure this out in your head, with a little bit of memorization and some practice.

The reason is the calendar days follow a regular pattern. Every 7 days cycles to the same day of the week. Because January 1, 2015 was a Thursday, that means the dates 1/8, 1/15, 1/22, and 1/29 must also fall on a Thursday. Similarly, the dates 1/3, 1/10, 1/17, 1/24, and 1/31 must be on a Saturday, because those dates are two days following the known date of Thursday January 1.

The mathematician John Conway devised a method called the Doomsday Rule that uses the pattern to allow easy mental calculation the day of the week for any date in a year. The method starts by memorizing dates in each month that fall on the same day of the week, known as a doomsday.

One set of dates are the doubles. In a year, the dates 4/4, 6/6, 8/8, 10/10, and 12/12 always fall on the same day of the week. The next set of dates falling on the same day of the week is 5/9, 7/11, 9/5, and 11/7. This can be memorized by the phrase "I work 9 to 5 at the store 7-11." We only need reference dates for the first three months now. In March, the date 3/14 is a doomsday (this is Pi Day because the constant π has the decimal expansion 3.14...). The last date in February is also on the same day, either 2/28 in a non-leap year or 2/29 in a leap year. Finally, the relevant date in January is 1/3 in a non-leap year and 1/4 in a leap year.

If you know the doomsday for a year, then you can find the day of the week for the remaining dates from the reference dates. Let's do some examples. We know January 3, 2015 falls on a Saturday. Which day does September 7, 2015 fall on? The closest reference date is 9/5, which also falls on a Saturday. Thus, September 7 is two days later on a Monday.

Now let's figure out Christmas December 25, 2015. The closest special date is 12/12, which falls on a Saturday. The 25th can be found by computing the difference in the number of days from the known date. That is, we calculate 25 – 12 = 13 days, which means 1 week, 6 days. Therefore, Christmas is one day before two weeks from Saturday, so it must be on a Friday.

The trick is to use the special dates as anchors and then figure out any other date relative to that. For practical purposes, you are almost always dealing with dates in the current year, so you only have to memorize the doomsday's day of the week for the current year.

But John Conway was not content with only the current year. He figured out how to generalize the rule to figure out the date in any year! This requires just a bit more memorization and calculation, but it allows you to figure out the day of the week for practically any date.

It will be useful to think about the days of the week in numerical codes. Here is a code with mnemonic tips, developed by Conway.

0 = Sunday = Nones-day

1 = Monday = One-day

2 = Tuesday = Twos-day

3 = Wednesday = Trebles-day

4 = Thursday = Fours-day

5 = Friday = Five-day

6 = Saturday = Six-a-day

We then need to memorize when the doomsday code for years ending with 00. These will serve as anchor days in our calculations.

1800 – 1899 = Friday = day code 5

1900 – 1999 = Wednesday = day code 3

2000 – 2099 = Tuesday = day code 2

2100 – 2199 = Sunday = day code 0

Now we can find the doomsday for any year from 1800 to 2199. Let's write the date as *CCYY* where *CC* is the century and *YY* is the last two digits. Here is the formula for the doomsday in a given year.

Doomsday code = [*YY* + rounddown(*YY*/4) + anchor(*CC*)] mod 7

Let me explain the terms in this formula. The term rounddown(*YY*/4) means to take the year *YY*, divide it by 4, and then round-down the result. For example, in 2015, we would have *YY* = 15 and so *YY*/4 = 15/4 = 3.75. We need to round-down the result, so the term is 3.

The term anchor(*CC*) means to take the anchor day code for the given century. In 2015, the century is the 2000s, and that anchor day is a Tuesday, with a day code of 2.

The final part is to take the entire result modulo 7. That means to only consider the remainder after 7. Equivalently, that means to reduce the number by multiples of 7 until the result is between 0 and 6. The reason we do this is we want to get a day code between 0 and 6, and after all, every 7 days corresponds to the same day of the week.

So let's use the formula on the year 2015. We will have:

Doomsday code = [*YY* + rounddown(*YY*/4) + anchor(*CC*)] mod 7

Doomsday code = [15 + rounddown(15/4) + anchor(20)] mod 7

Doomsday code = [15 + 3 + 2] mod 7

Doomsday code = [20] mod 7

Doomsday code = 6

So the doomsday in 2015 is a 6, or a Saturday. The key in knowing the doomsday for 2015 is we know all of the special dates fall on a Saturday as well, and other dates can be determined in relation to them.

Now let's put everything we've learned to solve a couple of examples. Let's find the day of the week for January 1, 1903. We first find the doomsday code, remembering the anchor for 1900s is a Wednesday (day code 3).

Doomsday code = [03 + rounddown(03/4) + anchor(19)] mod 7

Doomsday code = [3 + 0 + 3] mod 7

Doomsday code = 6

Therefore the Doomsday in 1903 was a Saturday. Since 1903 was not a leap year (leap years are divisible by 4), the closest memorable date is 1/3 which would also be a Saturday. Therefore January 1, 1903 fell two days earlier on a Thursday.

Let's do one more example of February 4, 2116. We first find the doomsday code, remembering the anchor for 2100s is a Sunday (day code 0).

Doomsday code = [16 + rounddown(16/4) + anchor(21)] mod 7

Doomsday code = [16 + 4 + 0] mod 7

Doomsday code = 20 mod 7

Doomsday code = 6

Therefore the Doomsday in 2116 falls on a Saturday. Since 2116 is a leap year (leap years are divisible by 4), the closest memorable date is the last day of February, 2/29, which would also be a Saturday. Then 2/4 would be 25 days earlier, which is 3 weeks and 4 days earlier. So four days before Saturday is a Tuesday, and therefore February 4, 2116 will be on a Tuesday.

I will mention there is also a special rule for century leap years. A year ending in 00 is only a leap year if it is also divisible by 400. So 2000 is a leap year, but 1800, 1900, and 2100 are not leap years.

Practice Problems

March 15, 2015

January 1, 2000

July 4, 1876

December 25, 2150

July 16, 1969

Proof

There are two parts we will justify. First, let's consider a specific year and the memorable dates. We claimed the same day of the week happens for the dates 4/4, 6/6, 8/8, 10/10, 12/12, 5/9, 9/5, 7/11, 11/7, and 3/14. This is a consequence that the same day of the week happens every 7 days, and each month has a specified number of dates. It just so happens that these dates are all multiples of 7 days apart, and hence they fall on the same day of the week. The dates for January and February depend on whether it is a leap year, because the extra day for the leap year changes which dates are multiples of 7 days apart. In a non-leap year, there is no extra day and the dates are 1/3 and 2/28. In a leap year, the extra day means the dates are 1/4 and 2/29.

The next part to justify is the formula for calculating the day of the week the doomsday falls on during a given year. Here is the formula.

Doomsday code = [*YY* + rounddown(*YY*/4) + anchor(*CC*)] mod 7

Why does this formula work? The formula starts with the doomsday for the 00 year of a given century, which is denoted anchor(*CC*). For any other year, *YY*, in that same century, we need to count the number of days that have passed. A normal year adds 365 days and a leap year adds 366 days. Since every 7 days is the same day of the week, we can reduce these numbers modulo 7 to find the offset. So a normal year adds 1 day, and a leap year adds 1 more day. So we can find the doomsday for the remaining years in a century by counting the number of years that have passed from 00, which is *YY*, and then adding in an extra day for each leap year, which is rounddown(*YY*/4). So this formula exactly accounts for the offset of each new year and leap year, and therefore it gives the doomsday for a year *YY* in a century *CC*.

Solutions to Practice Problems

March 15, 2015. The century code for 2000s is 2. So the doomsday in 2015 is calculated as (15 + rounddown(15/4) + 2) mod 7, which is then equal to 20 mod 7 = 6. So the doomsday falls on day code 6, a Saturday, and hence March 14 does as well. This means March 15, 2015 falls one day later on a Sunday.

January 1, 2000. The century code for 2000s is 2. So the doomsday in 2000 is calculated as (00 + rounddown(00/4) + 2) mod 7, which is then equal to 2 mod 7 = 2. So the doomsday falls on day code 2, a Tuesday, and hence in the leap year 2000 the date January 4 does as well. This means January 1, 2000 corresponds to 3 days earlier, on a Saturday.

July 4, 1876. The century code for 1800s is 5. So the doomsday in 1876 is calculated as (76 + rounddown(76/4) + 5) mod 7, which is then equal to 16 mod 7 = 2. So the doomsday falls on day code 2, a Tuesday, and hence July 11 (7/11) does as well. This means July 4, 1876 corresponds to one week earlier, also a Tuesday.

December 25, 2150. The century code for 2100s is 0. So the doomsday in 2150 is calculated as (50 + rounddown(50/4) + 0) mod 7, which is then equal to 13 mod 7 = 6. So the doomsday falls on day code 6, a Saturday, and hence December 12 (12/12) does as well. This means December 25, 2150 falls one week and six days later, on a Friday.

July 16, 1969. The century code for 1900s is 3. So the doomsday in 1969 is calculated as (69 + rounddown(69/4) + 3) mod 7, which is then equal to 89 mod 7 = 5. So the doomsday falls on day code 5, a Friday, and hence July 11 does as well. This means July 16, 2015 falls five days later on a Wednesday.

The Times Table 11 To 20

Can you multiply 12 by 14 easily? Problems up to 10 are often memorized when we learn the times table up to 10. This trick will help you figure out the times table from 11 to 20 by extending the times table up to 10.

There are four main steps. First, add the units digit of the second number to the first number. Second, multiply the result by 10. Third, multiply the units digits of both numbers. And finally, add the results from the second and third steps.

Let's do an example of 12 x 14. The number 14 has the units digit of 4. We add the units digit 4 to the number 12 to get 16. Then we multiply that by 10 to get 160. Third, we multiply the units digits of 2 and 4 to get 8. Finally, we add 160 and 8 to get 168.

Let's do another example of 15 by 18. We add 8 to 15 to get 23, and then we multiply that by 10 to get 230. We then multiply the units digits of 5 and 8 to get 40. Adding that to 230 gets an answer of 270.

Notice we would get the same answer even if we did 18 times 15. In this problem, we would add 5 to 18 to get 23, which we multiply by 10 to get 230. Then we multiply the units digits of 8 and 5 to get 40. Adding 230 and 40 once again results in the answer of 270.

This trick makes multiplying numbers between 11 and 20 an extension of the times tables from 1 to 10.

Practice Problems

11 x 16

12 x 17

14 x 19

15 x 13

16 x 16

Proof

Numbers between 11 and 19 can be written as $10 + x$ and $10 + y$, for x and y between 1 and 9. The product of two numbers can then be found as $(10 + x)(10 + y) = 100 + 10(x + y) + xy = 10[10 + (x + y)] + xy$.

The term $10 + x + y$ means to add the units digit of one number to the other number. Then that is multiplied by 10, after which the product of the units digits, xy, is added. This is the procedure described in this trick.

Solutions to Practice Problems

11 x 16. We add 6 to 11 to get 17, which is multiplied by 10 to get 170. Then the term 1 x 6 = 6 is added to get 176.

12 x 17. We add 7 to 12 to get 19, which is multiplied by 10 to get 190. Then the term 2 x 7 = 14 is added to get 204.

14 x 19. We add 9 to 14 to get 23, which is multiplied by 10 to get 230. Then the term 4 x 9 = 36 is added to get 266.

15 x 13. We add 3 to 15 to get 18, which is multiplied by 10 to get 180. Then the term 5 x 3 = 15 is added to get 195.

16 x 16. We add 6 to 16 to get 22, which is multiplied by 10 to get 220. Then the term 6 x 6 = 36 is added to get 256.

Multiply By 7, Then 11, Then 13

Ask a friend to pick a three-digit number. Tell them to multiply the number by 7, then 11, and then 13 on a calculator. Before they have the result, you will already know the answer.

Let's see how this works. Imagine your friend picks 123. You can instantly say the final result is 123,123. Or if your friend picks 479, you know the answer is 479,479.

The pattern is very simple so you can probably only do this trick once. The trick is that a number *ABC* multiplied by 7, then 11, then 13 will be *ABC*, *ABC* as a six-digit number.

Practice Problems

378 x 7 x 11 x 13

926 x 7 x 11 x 13

Proof

Note that 7 x 11 x 13 = 1,001. When you multiply *ABC* by 1,001, that is the same as multiplying by (1,000 + 1). Multiplying by 1,000 shifts the number *ABC* by three spots to get *ABC*,000. Then we multiply by 1 and add the result, which means adding *ABC* to the number again. The final result is *ABC*, *ABC* as a six-digit number.

Solutions to Practice Problems

378 x 7 x 11 x 13. The answer is the three-digit number 378 repeated twice. The result is 378,378.

926 x 7 x 11 x 13. The answer is the three-digit number 926 repeated twice. The result is 926,926.

Regroup Factors

Can you multiply 25 by 80? Sometimes we can re-group factors of the numbers to make the problem simpler. For example, we can think about 80 as 4 x 20, so we have 25 x 80 = 25 x (4 x 20). We can then write this as (25 x 4) x 20 = 100 x 20 = 2,000. So we can also conclude 25 by 80 is equal to 2,000.

The trick of regrouping factors is about shifting around factors from both numbers to get more convenient numbers. Which numbers are "convenient?"

Mostly we look for ways to create factors of 10, because it is easy to multiply by 10. Here are some more problems we can use the method of regrouping factors.

Multiply 5 by 46. The trick is that 5 x 2 = 10, and since 46 is an even number it has a factor of 2. Therefore 5 x 46 = 5 x (2 x 23), and we can write this as (5 x 2) x 23 = 10 x 23. This last problem is easy as multiplying by 10 is the same as "adding a zero" to the end of the number. Thus, 5 x 46 = 230.

Multiply 15 by 66. We again want to find a factor of 10. We can factor a 5 from the number 15, and since 66 is even we can factor a 2 from it. Therefore, we have 15 x 66 = (3 x 5) x 66 = (3 x 5) x (2 x 33). Now we write 3 x (5 x 2) x 33 = 3 x 10 x 33 = (3 x 33) x 10 = 99 x 10 = 990. This might seem like a daunting sequence of steps, but it will become natural after you start looking for ways to regroup factors.

Regrouping factors is a method to create an equivalent multiplication problem that is easier to calculate. This mostly comes in handy when you have numbers that end in 5 multiplied by even numbers, which have a factor of 2. You can combine the factors of 5 and 2 to get a factor of 10, which makes the problem easier.

Or you might have multiple factors of 5 and 2. For example, let's do 125 times 44. The number 125 is 5 x 25 and the number 44 is 4 x 11. So we can calculate 125(44) = (5 x 25)(4 x 11) = 5(25 x 4)11, and then we finish the steps as 5(100)11 = 500(11) = 5,500.

Regrouping factors is a good mental math trick if you can pull it off. But it is also useful when you are doing problems by hand.

Practice Problems

25 x 16

75 x 6

150 x 80

35 x 12

45 x 8

Proof

Re-grouping works because multiplication is associative and commutative. That is, we can multiply numbers, or factors of numbers, in a re-grouped order and get the same result.

Solutions to Practice Problems

25 x 16. We re-group this problem as 25 x 16 = 25 x (4 x 4), which is then equal to (25 x 4) x 4 = 100 x 4 = 400.

75 x 6. We re-group this problem as 75 x 6 = 75 x (2 x 3), which is then equal to (75 x 2) x 3 = 150 x 3 = 450.

150 x 80. We re-group this problem as 150 x 80 = 150 x (4 x 20), which is then equal to (150 x 4) x 20 = 600 x 20 = 12,000.

35 x 12. We re-group this problem as 35 x 12 = 35 x (2 x 6), which is then equal to (35 x 2) x 6 = 70 x 6 = 420.

45 x 8. We re-group this problem as 45 x 8 = 45 x (2 x 4), which is then equal to (45 x 2) x 4 = 90 x 4 = 360.

Split Up A Multiplication

What is 93 times 7? This problem is a lot easier to solve in your head if you split it up into two easier problems. We can think about 93 = 90 + 3, and then we can do (90 + 3) x 7 = 90 x 7 + 3 x 7. While this might look more difficult, it is actually easier to do because the problem involves adding up the result of two easier tasks. Since we have 90 x 7 = 630 and 3 x 7 = 21, the answer is the sum of 651.

Let's try another problem. What is 93 x 17? Multiplying two-digit numbers in your head would be very hard to do the traditional way, as you have to remember to carry over. So we can split up the problem as the problem 93 x 17 = 93 x (20 – 3) = 93 x 20 – 93 x 3. The first part of the expression 93 x 20 = 1,860 is easy. But then to do 93 x 3, we again split it up as (90 + 3) x 3 = 90 x 3 + 3 x 3 = 270 + 9 = 279. So we subtract 279 from 1,860 to get 1,581. (Note the subtraction could be split up as well to 1,860 – 279 = 1,860 – 300 + 21 = 1,560 + 21 = 1,581.)

In general, try to re-write one of the numbers as a difference from a multiple of 10, 100, etc. and then you can break the problem down into easier parts.

Practice Problems

97 x 3

101 x 12

11 x 26

198 x 21

17 x 212

Proof

Splitting up a multiplication works because of the distributive property of multiplication, which states $(x + y) z = xz + yz$. So we can break up a number into two parts x and y and then add up the product of each part with z.

Solutions to Practice Problems

97 x 3. This problem can be split up as 97 x 3 = (100 – 3) x 3, which is then equal to 100(3) – (3)(3) = 300 – 9 = 291.

101 x 12. This problem can be split up as 101 x 12 = (100 + 1) x 12, which is then equal to 100(12) + (1)(12) = 1,200 + 12 = 1,212.

11 x 26. This problem can be split up as 11 x 26 = (10 + 1) x 26, which is then equal to 10(26) + (1)(26) = 260 + 26 = 286.

198 x 21. This problem can be split up as 198 x 21 = (200 – 2) x 21, which is then equal to 200(21) – (2)(21) = 4,200 – 42 = 4,158.

17 x 212. This problem can be split up as 17 x 212 = 17 x (200 + 12), which is then equal to 17(200) + 17(12) = 3,400 + 17(10 + 2). We can continue evaluating 3,400 + 17(10) + 17(2) = 3,400 + 170 + 34, and finally we have 3,570 + 34 = 3,604.

Divide By 4, 8, Etc.

What is 256 divided by 4? Instead of working it out by long division, you can divide by 4 by halving two times. So 256 halved is 128, which we then halve again to get 64.

Similarly, you can divide by 8 by halving three times. For example, let's do 1,024 divided by 8. We can halve 1,024 to get 512, then we halve that to get 256, and we halve one more time to get 128.

In general, you can divide by 2^n if you repeatedly halve a total of n times.

Practice Problems

$72 \div 4$

$135 \div 4$

$100 \div 8$

$1{,}234 \div 8$

$2{,}082 \div 16$

Proof

Dividing a number x by 4 is equivalent to dividing x by 2^2. So we have $x/4 = x/(2^2) = (x/2)(1/2) = (x/2)/2$. In other words, we can divide by 4 by dividing by 2 twice.

Similarly, dividing a number x by 2^n can be expressed as dividing by 2 a total of n times. So we have $x/(2^n) = (x/2)(1/2)(1/2)...(1/2)$, which is equal to $(x/2)/2 \ldots / 2$, where we are dividing by 2 a total of n times.

Solutions to Practice Problems

$72 \div 4$. We need to halve two times to divide by $4 = 2^2$. We halve 72 to get 36, which is then halved again to get 18.

$135 \div 4$. We need to halve two times to divide by $4 = 2^2$. We halve 135 to get 67.5, which is then halved again to get 33.75.

100 ÷ 8. We need to halve three times to divide by $8 = 2^3$. We halve 100 to get 50, then halve that to get 25, and finally halve that to get 12.5.

1,234 ÷ 8. We need to halve three times to divide by $8 = 2^3$. We halve 1,234 to get 617, then we have again to get 308.5, and we halve a third time to get 154.25.

2,082 ÷ 16. We need to halve four times to divide by $16 = 2^4$. We halve 2,082 to get 1,041, then we have again to get 520.5, then we halve a third time to get 260.25, and then we halve a fourth time to get 130.125.

Divide By 5, 50, Etc.

What is 350 divided by 5?

A simple way to find the answer is to note that 1/5 = 2/10. So dividing by 5 is the same as dividing by 10 and then multiplying by 2. So 350 divided by 10 is 35, which we then double to get 70.

This trick is useful for any number that starts with a 5 and has only 0s after that. For example, let's divide 1,235 by 50. We first divide 1,235 by 100 to get 12.35. Then we double that to get 24.70.

In general, to divide by 50..0, which has N zeros, you can divide by 100..0, which has $N + 1$ zeros, and then double the result. Alternately, you can double the result and then divide by 100..0, which has $N + 1$ zeros.

You can use a similar trick to divide by 25, 250, etc. The key is to observe that 1/25 = 4/100. So you can divide by 25 by dividing by 100 and then doubling two times. So what is 1,235 divided by 25? We first divide by 100 to get 12.35. Then we double to get 24.70, and then we double again to get 49.40.

In general, to divide by 250..0, which has N zeros, you can divide by 100..0, which has $N + 2$ zeros, and then quadruple the result by doubling two times. Alternately, you can do the quadrupling first and then divide by 100..0.

Practice Problems

315 ÷ 50

1,210 ÷ 50

1,324 ÷ 50

11,434 ÷ 500

22,082 ÷ 5,000

Proof

Since $50 = 100/2$, dividing x by 50 means $x/50 = x/(100/2) = (2x)/100$. In other words, we can divide by 50 by doubling the number and then dividing by 100. Equivalently, we can divide by 100 and then double the number.

Similarly, dividing by $5(10^n)$ is the same as doubling the number and then dividing by 10^{n+1}. Therefore $x/[5(10^n)] = x/(10^{n+1}/2) = (2x)/10^{n+1}$.

Dividing a number x by $25(10^n)$ can be expressed as quadrupling the number and then dividing by 10^{n+2}. The proof is $x/[25(10^n)]$ can be written as $x/(10^{n+2}/4) = (4x)/10^{n+2}$.

Solutions to Practice Problems

315 ÷ 50. We divide by 100 to get 3.15 and then double that to get 6.3.

1,210 ÷ 50. We divide by 100 to get 12.1 and then double to get 24.2.

1,324 ÷ 50. We divide by 100 to get 13.24 and then double to get 26.48.

11,434 ÷ 500. We divide by 1,000 to get 11.434 and then double to get 22.868.

22,082 ÷ 5,000. We divide by 10,000 to get 22.082 and then double to get 4.4164.

Part II: Squaring Numbers

Square A Number Ending In 5

Can you square 35 in your head? What about 65 squared? It is very easy to find the square of a number ending in 5.

There are two steps. First, multiply the tens digit by one more than itself. Second, append the number 25 to the result from step 1.

For example, let's calculate 35^2. The tens digit of 35 is 3. So we multiply 3 by one more than itself, which is 4, to get 3 x 4 = 12. The second step is to append 25 to the 12, which means the answer is 1,225.

For another example, let's do 65^2. The tens digit is 6 and one more than that is 7. So we multiply 6 by 7 to get 42. We then append 25 to the end of the number to get an answer of 4,225.

This trick actually works for three-digit numbers and larger numbers too. The only modification is the "tens" digit is the number resulting from deleting the 5 from the number. So in the number 125, make the "tens" digit the number by deleting the 5, which is the number 12.

For example, let's say we want to square 105. For the first step, we delete the 5 and we are left with the number 10. We multiply 10 by one more than itself, 11, to get 110. Then we append 25 to get 11,025 as the answer.

Take some time to learn and practice this rule because some of the later tricks in the book will build upon this trick.

Practice Problems

15^2

45^2

75^2

125^2

255^2

Proof

Squaring a number ending in 5 means to multiply $(10x + 5)$ times itself. By algebra, we have $(10x + 5)(10x + 5) = 100x^2 + 100x + 25$, which is then equal to $100x(x + 1) + 25$. The term 25 gives the last two digits of 25, and the term $100x(x + 1)$ indicates to multiply the number x times one more than itself, $x + 1$, and then shift over two places so these are the leading digits of the answer.

Solutions to Practice Problems

15^2. We find the answer starts as 1 x 2 = 2 and then ends with 25. So the answer is $15^2 = 225$.

45^2. We find the answer starts as 4 x 5 = 20 and then ends with 25. So the answer is $45^2 = 2{,}025$.

75^2. We find the answer starts as 7 x 8 = 56 and then ends with 25. So the answer is $75^2 = 5{,}625$.

125^2. We find the answer starts as 12 x 13. There are many ways to solve this. We will re-write 12 x 13 = 12 x (12 + 1). Then we can solve the problem as $12^2 + 12 = 144 + 12 = 156$. Those are the first three digits of the answer, and we append the digits 25. So the answer is $125^2 = 15{,}625$.

255^2. We find the answer starts as 25 x 26 = $25^2 + 25$. We can find 25^2 by the same trick. So we do 2 x 3 = 6 and append 25 to get 625. Thus we have 25 x 26 = $25^2 + 25 = 625 + 25 = 650$. The answer then ends with the digits 25. So the answer is $255^2 = 65{,}025$.

Square A Number Ending In 9 Or 4

It is easy to square a number ending in 0 because a number ending in 0 is a multiple of 10. The rule is to square the leading digits and then double the number of 0s. For example, 30^2 can be calculated by squaring 3, which is 9, and then doubling the number of zeros to 00 to get the answer 900. Similarly, 80^2 is equal to $8^2 = 64$ followed by 00, or 6,400.

It is also relatively easy to square a number ending in 5, as already explained. The rule is to multiply the leading digits by a number one more than itself and then append 25 to the result. For example, 35^2 can be calculated by multiplying 3 by 4, which is 12, and then appending 25 to get the answer 1,225. Similarly, 85^2 is equal to 8 x 9 = 72 followed by 25, or 7,225.

Numbers ending in 9 are one less than numbers ending in 0. Similarly, numbers ending in 4 are one less than numbers ending in 5. As such, we can square numbers ending in 9 or 4 by squaring numbers one more than them and then adjusting the result.

Here is the rule. Let's say the number ends in a 9 or 4. We can calculate by squaring one more than the number, and then we subtract the original number and one more than the original number.

Let's do some examples. Let's calculate 39^2. One more than the number is 40, and $40^2 = 1{,}600$. We then subtract the original number, 39, and one more than the original number, 40. So we take 1,600 and subtract 39 and 40. The end result is $39^2 = 1{,}521 = 1{,}600 - 39 - 40$.

Let's calculate 74^2. One more than the number is 75, and $75^2 = 5{,}625$ by the rule of squaring a number ending in 5. We then subtract the original number, 74, and one more than the original number, 75. So we take 5,625 and subtract 74 and 75. The end result is $74^2 = 5{,}476$.

To summarize, you can square a number ending in 9 or 4 by squaring the number one above it, and then subtracting out the original number and one more than the original number. You will find it easier to square the number one more—as it will be a number ending in 0 or 5—and then you can adjust the result by subtracting the original number and one more than the original number.

Practice Problems

14^2

49^2

84^2

129^2

254^2

Proof

Here is why the rule works. For any number x, you can obtain x^2 by squaring one more than x and then subtracting x and $x + 1$. The algebraic proof is $(x + 1)^2 - x - (x + 1) = (x^2 + 2x + 1) - 2x - 1 = x^2$.

Solutions to Practice Problems

14^2. We need to calculate 15^2 first. This is found by the rule of squaring a number ending in 5. So we calculate 1 x 2 = 2 and then append the digits 25, so $15^2 = 225$. Then we have $14^2 = 15^2 - 14 - 15 = 225 - 29 = 196$.

49^2. We calculate $50^2 = 2{,}500$ first. Then we subtract 49 and 50 to get the result $49^2 = 2{,}401$.

84^2. We need to calculate 85^2 first. This is found by the rule of squaring a number ending in 5. So we calculate 8 x 9 = 72 and then append the digits 25, so $85^2 = 7{,}225$. Then we have $84^2 = 85^2 - 84 - 85 = 7{,}056$.

129^2. We calculate $130^2 = 16{,}900$ first. Then we subtract 130 and 129 to get $129^2 = 16{,}900 - 260 + 1 = 16{,}641$.

254^2. We need to calculate 255^2 first. This is found by the rule of squaring a number ending in 5. So we calculate 25 x 26 = $25^2 + 25$. We can find 25^2 by the same trick. So we do 2 x 3 = 6 and append 25 to get 625. Thus 25 x 26 = $25^2 + 25 = 625 + 25 = 650$. The answer then ends with 25, so $255^2 = 65{,}025$. Then we have $254^2 = 255^2 - 255 - 254$, which is equal to $65{,}025 - 510 + 1 = 64{,}516$.

Square A Number Ending In 1 Or 6

Numbers ending in 1 are one more than numbers ending in 0. Similarly, numbers ending in 6 are one more than numbers ending in 5. The rule for numbers ending in 1 or 6 is very similar to the one for numbers ending in 9 or 4.

Here is the rule. Let's say the number ends in a 1 or 6. We calculate by squaring one less than the number, and then we add back the original number and one less than the original number.

Let's do some examples. Let's calculate 21^2. One less than the number is 20, and $20^2 = 400$. We then add the original number, 21, and one less than the original number, 20, for an adjustment of 41. The end result is $21^2 = 441$.

Let's calculate 86^2. One less than the number is 85, and $85^2 = 7{,}225$ (by the rule of squaring a number ending in 5). We then add the original number, 86, and one less than the original number 85, for an adjustment of 171. The end result is $86^2 = 7{,}396$.

How can you remember the rule for a number ending in 4 or 9 versus the rule for a number ending in 1 or 6? The key is to think about how the original number compares to the convenient number. When the original number is smaller (like in 4 or 9), you need to adjust the square by subtracting the convenient number and the original number. When the original number is larger (like in 1 or 6), you need to adjust by adding the convenient number and the original number.

Practice Problems

16^2

41^2

76^2

121^2

256^2

Proof

Here is why the rule works. For any number x, you can obtain x^2 by squaring one less than x and then adding x and $x - 1$. The algebraic proof is $(x - 1)^2 + x + (x - 1) = (x^2 - 2x + 1) + (2x - 1) = x^2$.

Solutions to Practice Problems

16^2. We need to calculate 15^2 first. This is found by the rule of squaring a number ending in 5. So we calculate 1 x 2 = 2 and then append the digits 25, so $15^2 = 225$. Then we have $16^2 = 15^2 + 15 + 16 = 225 + 31 = 256$.

41^2. We calculate $40^2 = 1,600$ first. Then we add 40 and 41 to get the answer $41^2 = 1,681$.

76^2. We need to calculate 75^2 first. This is found by the rule of squaring a number ending in 5. So we calculate 7 x 8 = 56 and then append the digits 25, so $75^2 = 5,625$. Then we have $74^2 = 75^2 + 75 + 76$, which then equals $5,625 + 151 = 5,776$.

121^2. We calculate $120^2 = 14,400$ first. Then we add 120 and 121 to get $121^2 = 14,400 + 241 = 14,641$.

256^2. We need to calculate 255^2 first. This is found by the rule of squaring a number ending in 5. So we calculate 25 x 26 = $25^2 + 25$. We can find 25^2 by the same trick. So we do 2 x 3 = 6 and append 25 to get 625. Thus 25 x 26 = $25^2 + 25 = 625 + 25 = 650$. The answer then ends with 25, so $255^2 = 65,025$. Then we have $256^2 = 255^2 + 255 + 256$, which is the result of $65,025 + 511 = 65,536$.

Square A Number Ending In 8 Or 3, Or A Number Ending In 7 Or 2

For completeness, we will present similar rules for numbers ending in 8 or 3, and 7 or 2. These rules are not as easy to remember, but they are shown to illustrate how you can extend the previous rules for the remaining digits.

Here is the rule for 8 or 3. The convenient number is 2 more than the original number. We can calculate the square by calculating the square of the convenient number, subtracting 4 times the convenient number, and then adding back 4.

Let's do some examples. Let's calculate 38^2. The convenient number is 2 more than the original number, 40. So we have $40^2 = 1{,}600$. We then quadruple the convenient number 40 to get 160. Subtracting 160 from 1,600 results in 1,440. Finally, we add 4 to get $1{,}444 = 38^2$.

Let's calculate 73^2. Two more than the number is the convenient number of 75, and $75^2 = 5{,}625$ by the rule of squaring a number ending in 5. We then quadruple 75 to get 300. Subtracting 300 from 5,625 results in 5,325. Finally, adding 4 means $73^2 = 5{,}329$.

The rule for 7 or 2 is similar, but the nearby convenient number is two less than the number ending in 7 or 2. So we can calculate the square by calculating the square of the convenient number, adding 4 times the convenient number, and then adding back 4.

Let's do some examples. Let's calculate 62^2. The convenient number is two less than 62, which is 60, and $60^2 = 3{,}600$. We then quadruple the convenient number 60 to get 240. Adding 240 to 3,600 results in 3,840. Finally we add 4 to get $3{,}844 = 38^2$.

Let's calculate 47^2. Two less than the number is 45, and $45^2 = 2{,}025$ by the rule of squaring a number ending in 5. We then quadruple 45, which is 180. Adding 180 to 2,025 results in 2,205, and then adding 4 means $47^2 = 2{,}209$.

In the cases of 8 or 3, as well as 7 or 2, you always square the convenient number and then add 4 at the end. If the convenient number is larger than the original number (8 or 3), then you need to subtract quadruple the convenient number. If the convenient number is smaller than the original number (7 or 2), then you need to add quadruple the convenient number.

Practice Problems

18^2

43^2

77^2

122^2

252^2

Proof

Here is a proof for the rule of 8 or 3. For any number x, you can obtain x^2 by squaring $x + 2$, subtracting 4 times $x + 2$, and then adding 4. The algebraic proof is $(x + 2)^2 - 4(x + 2) + 4 = (x^2 + 4x + 4) - 4x - 8 + 4 = x^2$.

Here is a proof for the rule of 7 or 2. For any number x, you can obtain x^2 by squaring $x - 2$, adding 4 times $x - 2$, and then adding 4. The algebraic proof is $(x - 2)^2 + 4(x - 2) + 4 = (x^2 - 4x + 4) + 4x - 8 + 4 = x^2$.

Solutions to Practice Problems

18^2. We first calculate $20^2 = 400$. Then we subtract 4 x 20 = 80 and add 4, so we have $18^2 = 400 - 80 + 4 = 324$.

43^2. We calculate 45^2 first by the rule of squaring a number ending in 5. This is $45^2 = 2{,}025$. Then we subtract 4 x 45 = 2 x 90 = 180 and then add 4. So we have $43^2 = 2{,}025 - 180 + 4 = 1{,}849$.

77^2. We need to calculate 75^2 first. This is found by the rule of squaring a number ending in 5. So we calculate 7 x 8 = 56 and then append the digits 25, so $75^2 = 5{,}625$. Then we need to add 4 x 75 = 2 x 150 = 300 and add 4. So we have $77^2 = 75^2 + 300 + 4 = 5{,}625 + 304 = 5{,}929$.

122^2. We calculate $120^2 = 14{,}400$ first. Then we add 4 x 120 = 480 and 4 to get $122^2 = 14{,}400 + 480 + 4 = 14{,}884$.

252^2. We need to calculate 250^2 first which is 62,500 (this can be found because $250^2 = (25 \text{ x } 10)^2 = 25^2 \text{ x } 10^2 = 25^2 \text{ x } 100$, and 25^2 can be found by the method for squaring a number ending in 5). Then we need to add $4 \text{ x } 250 = 1{,}000$ and 4. Thus $252^2 = 250^2 + 1{,}000 + 4$, and finally this gets to the answer of $62{,}500 + 1{,}004 = 63{,}504$.

Square A Two-Digit Number General Procedure

There is an alternate method to squaring any two-digit number if you are comfortable doing a few calculations in your head involving carrying over. Let's do an example and then explain the procedure.

What is the square of 23? We add the second digit to the original number. So we add 3 to 23 to get 26. Then we multiply this by the first digit of 2 to get 52, which we append a 0 to get 520. We finally add in the square of the second digit 3, which is 9. The result is 529.

The procedure breaks down squaring the number into the following easier calculations: add the second digit to the original number, multiply by the first digit, multiply by 10, and then finally add the square of the second digit.

Let's try another example of the square of 72. We add 2 to 72 to get 74, which is then multiplied by 7 to get 518. We multiply by 10 to get 5,180. Finally we add the square of 2, which is 4, to get 5,184.

Practice Problems

28^2

34^2

17^2

86^2

91^2

Proof

A two-digit number x can be expressed as $x = 10a + b$. From algebra, we have $x^2 = (10a + b)^2 = 100a + 20ab + b^2 = 10a(10a + 2b) + b^2$, which equals $10a(x + b) + b^2$. The term $x + b$ means to add the second digit of b to the number x. That is multiplied by the first digit of a, which is further multiplied by 10. Finally the square of second digit b is added.

Solutions to Practice Problems

28^2. We add 8 to 28 to get 36, then that is multiplied by 2 to get 72, and that is further multiplied by 10 to get 720. Then we add $8^2 = 64$ to get the answer 784.

34^2. We add 4 to 34 to get 38, then that is multiplied by 3 to get 114, and that is further multiplied by 10 to get 1,140. Then we add $4^2 = 16$ to get the answer 1,156.

17^2. We add 7 to 17 to get 24. Multiplying by 1 is still 24, and that is further multiplied by 10 to get 240. Then we add $7^2 = 49$ to get the answer 289.

86^2. We add 6 to 86 to get 92. Now we need to multiply 92 by 8. We can instead do 92 times $(10 - 2)$ to get $920 - 184 = 736$. That is multiplied by 10 to get 7,360. Then we add $6^2 = 36$ to get the answer 7,396.

91^2. We add 1 to 91 to get 92, then that is multiplied by 9 to get 828, and further multiplied by 10 to get 8,280. Then we add $1^2 = 1$ to get the answer 8,281.

Square A Number Between 41 And 59

How would you square 54? Many methods have been explained already. There is in fact a very quick way to do this problem that is worth learning.

The first step is to consider the difference between 54 and 50. The difference is +4. The last two digits of our answer are the square of the difference. Since 4 squared is equal to 16, the last two digits of our answer are 16.

The first two digits of the answer will be half of 50, which is 25, plus the difference of +4. Since 25 + 4 = 29, the first two digits of the answer are 29.

Putting this together, the answer is 2,916.

We can similarly square any number from 51 to 59 in the following process. The first two digits are equal to 25 plus the difference, and the last two digits are equal to the difference squared. For example, 58 has a difference of +8. The first two digits will be 25 + 8 = 33 and the last two digits will be $8^2 = 64$. So the answer is 3,364.

How would we find the square of 47? The process is almost the same. The only change that the number 47 is less than 50. Therefore, we need to subtract the difference from 25.

Here's how we can square 47. We notice the number 47 is 3 less than 50. The first two digits of our answer are 25 – 3 = 22. The last two digits are the square of 3, which we write as a two-digit number 09. Thus we have 2,209 as the answer.

You can also apply this rule to remember the square of 50. This number has a difference of 0 from 50, and so the first two digits are 25 + 0 = 25, and the last two digits are the square of 0, which is 00 when written as a two-digit number. Thus the answer is 2,500.

To summarize, you can square a number from 41 to 59 as follows. Adjust the number 25 by the difference from 50 to get the first two digits of the answer. Then, square the difference, and write that as a two-digit number, to get the last two digits of the answer.

Practice Problems

44^2

57^2

46^2

43^2

52^2

Proof

For any number x, $(50 + x)^2 = 2{,}500 + 100x + x^2 = 100(25 + x) + x^2$. The term $25 + x$ means to adjust the number 25 by x, which is the difference from 50. This is multiplied by 100 to become the first two digits of the answer. Then the square of the difference is added, and those two digits are the last two digits of the answer.

Solutions to Practice Problems

44^2. The number 44 is 6 less than 50. So we subtract 6 from 25 to get 19, and then append $6^2 = 36$. The result is 1,936.

57^2. The number 57 is 7 more than 50. So we add 7 to 25 to get 32, and then append $7^2 = 49$. The result is 3,249.

46^2. The number 46 is 4 less than 50. So we subtract 4 from 25 to get 21, and then append $4^2 = 16$. The result is 2,116.

43^2. The number 43 is 7 less than 50. So we subtract 7 from 25 to get 18, and then append $7^2 = 49$. The result is 1,849.

52^2. The number 52 is 2 more than 50. So we add 2 to 25 to get 27, and then append $2^2 = 04$ (remembering to write the square as a two-digit number!). The result is 2,704.

Square A Number Around 500

How would you square 504? There is a general rule similar to the one we just explained for squaring a number around 50.

The first step is to consider the difference between 504 and 500. The difference is +4. The last *three* digits of our answer—since 500 is a three-digit number—will be the square of the difference. Since 4 squared is equal to 16, the last three digits of our answer are 016.

The first *three* digits of the answer will half of 500, which is 250, plus the difference of 4. Since $250 + 4 = 254$, the first three digits of the answer are 254. Putting this together, the answer is 254,016.

We can similarly square any number above 500 by the same process. The first three digits of the answer will be 250 plus the difference, and the last three digits will be the difference squared, written as a three-digit number. For example, 506^2 can be calculated as $250 + 6 = 256$ (the first three digits) and $6^2 = 36 = 036$ (the last three digits). So the answer is 256,036.

We can similarly find the squares of numbers less than 500. We just need to remember to subtract the difference from 250 since these numbers are less than 500. So let's square 498. The difference of this number is 2. The first three digits will be $250 - 2 = 248$, and the last three digits will be 2 squared, which is 004 when written as a three-digit number. So 498 squared is equal to 248,004.

You can generalize this process to find the square near any number that starts with a 5 and ends with 0s. For example, let's do 5,000. Since 5,000 is a four-digit number, the first four digits will be 2,500 plus/minus the difference, and the final four digits will be the difference squared. To square numbers near 50,000, it's the same rule except we now have a five digit number. Therefore the first five digits will be 25,000 plus/minus the difference, and the final five digits will be the difference squared.

You can see how the pattern continues. The number with 25 is always half of the original number. So with 50 the number is 25, then with 500 it is 250, with 5,000 it is 2,500, and so on.

Practice Problems

505^2

507^2

496^2

493^2

$5,002^2$

Proof

A number near 50, 500, 5,000, or so on, can be written as $5(10)^k + x$. To square the number, we have $(5(10)^k + x)^2 = 25(10)^{2k} + (10)^{k+1}x + x^2$. This equals $(10)^{k+1}(25(10)^{k-1} + x) + x^2$. In other words, we add the difference x to the number 25 with appropriate number of zeros to become the leading digits of the answer. Then we add the x^2 term, written with the appropriate number of digits, to become the ending digits of the answer.

Solutions to Practice Problems

505^2. The number 505 is 5 more than 500. So we add 5 to 250 to get 255, and then append $5^2 = 025$ (remembering to write as a three-digit number). The result is 255,025.

507^2. The number 507 is 7 more than 500. So we add 7 to 250 to get 257, and then append $7^2 = 049$. The result is 257,049.

496^2. The number 496 is 4 less than 500. So we subtract 4 from 250 to get 246, and then append $4^2 = 016$. The result is 246,016.

493^2. The number 493 is 7 less than 500. So we subtract 7 from 250 to get 243, and then append $7^2 = 049$. The result is 243,049.

$5,002^2$. The number 5002 is 2 more than 5,000. So we add 2 to 2,500 to get 2,502, and then append $2^2 = 0004$ (with four-digit numbers we need this square to be written as a four-digit number). The result is 25,020,004.

Square A Two-Digit Number Faster

There is a quicker way to square any two-digit number, if you first memorize the squares up to 25.

$1^2 = 1$, $2^2 = 4$, $3^2 = 9$, $4^2 = 16$, $5^2 = 25$

$6^2 = 36$, $7^2 = 49$, $8^2 = 64$, $9^2 = 81$, $10^2 = 100$

$11^2 = 121$, $12^2 = 144$, $13^2 = 169$, $14^2 = 196$, $15^2 = 225$

$16^2 = 256$, $17^2 = 289$, $18^2 = 324$, $19^2 = 361$, $20^2 = 400$

$21^2 = 441$, $22^2 = 484$, $23^2 = 529$, $24^2 = 576$, $25^2 = 625$

The squares from 26 to 99 can be calculated from these memorized values.

For the numbers between 26 and 75, the method is to use the rule for squaring numbers around 50. We find the difference from 50, and we either subtract that (for numbers less than 50) or we add that (for numbers larger than 50) from the number 25. These are the first two digits of the answer. Then we square the difference and add that as the last two digits. If the square of the difference is three digits, then we have to carry over to the first two digits.

Let's do an example of 38^2. This number is 12 *less* than 50, so we *subtract* 12 from 25 to get 13. These are the first two digits of the answer. Then we square the difference of 12 to get 144. The 44 are the last two digits, and the 1 carries over to the 13 to become 14. Therefore the answer is 1,444.

Let's do another example of 73^2. This number is 23 *more* than 50, so we *add* 23 to 25 to get 48. These are the first two digits of the answer. Then we square the difference of 23 to get 529 (this is why the squares up to 25 have to be memorized). The 29 are the last two digits, and the 5 carries over to the 48 to become 53. Therefore the answer is 5,329.

For numbers 76 to 99, we can look at the difference from 100. The procedure is to subtract double the difference from 100, and then add the square of the difference, with carry-over if the square is three digits or more.

For example, let's do 87^2. The difference from 100 is 13, so we subtract double the difference of 26 from 100 to get 74. These are the first two digits. Then we square the difference of 13 to get 169. Since this is a three-digit result, we carry over the 1 to the 74 to make 75. So the answer is 7,569.

For another example, let's do 79^2. The difference from 100 is 21, so we subtract double the difference of 42 from 100 to get 58. These are the first two digits. Then we square the difference of 21 to get 441. Since this is a three-digit result, we carry over the 4 to the 58 to make 62. So the answer is 6,241.

In fact, you can quickly adapt this trick to do squares between 101 and 125. The rule is to add double the difference to 100, and then add the square of the difference.

For instance, 112^2 can be calculated as $100 + 2(12) = 124$ as the first three digits, and then $12^2 = 144$. We carry over the 1 to make 124 into 125. So the answer is 12,544.

Practice Problems

36^2

29^2

67^2

91^2

117^2

Proof

The rule for numbers 26 to 75 works for the same reason as squaring a number between 41 and 59, which was already proven.

For numbers between 76 and 100, the rule works because for any number x, we have $(100 - x)^2 = 10{,}000 - 200x + x^2 = 100(100 - 2x) + x^2$. That is, we subtract the double difference x from 100, shift over by 2 decimal places, and then add x^2.

For numbers 101 to 125, the rule is to add double the difference to 100, and then add the square of the difference. This works because for any number x, we have $(100 + x)^2 = 10{,}000 + 200x + x^2 = 100(100 + 2x) + x^2$.

Solutions to Practice Problems

36^2. The number 36 is 14 less than 50. So we subtract that from 25 to get 11. Then we add $14^2 = 196$. The digits 96 become the last two digits, and the 1 carries over to the 11 to make 12. So the answer is 1,296.

29^2. The number 29 is 21 less than 50. So we subtract that from 25 to get 4. Then we add $21^2 = 441$. The digits 41 become the last two digits, and the hundreds digit 4 carries over to the 4 to make 8. So the answer is 841.

67^2. The number 67 is 17 more than 50. So we add that to 25 to get 42. Then we add $17^2 = 289$. The digits 89 become the last two digits, and the hundreds digit 2 carries over to the 42 to make 44. So the answer is 4,489.

91^2. The number 91 is 9 less than 100. So we subtract double the amount, 18, from 100 to get 82. Then we add $9^2 = 81$. The result is 8,281.

117^2. The number 117 is 17 more than 100. So we add double the amount, 34, to 100 to get 134. Then we add $17^2 = 289$. The hundreds digit of 2 carries over to 134 to make 136. The digits 89 are the last two digits of the answer. The result is 13,689.

Square 34, 334, Etc. Or Square 67, 667, Etc.

This is a fairly useless math trick, but there is a mathematical reason for why this trick was discovered, which is why this trick is mentioned.

There is a pattern to squaring numbers starting with a series of 3s and ending in a 4. Here are some examples.

$34^2 = 1,156$

$334^2 = 111,556$

$3,334^2 = 11,115,556$

For a number starting with N digits of a 3 and ending in a 4, the square of that number will have $N + 1$ digits of 1, then N digits of 3, and the last digit will be a 6.

Similarly, there is a pattern to squaring numbers starting with a series of 6s and ending in a 7. Here are some examples.

$67^2 = 4,489$

$667^2 = 444,889$

$6,667^2 = 44,448,889$

For a number starting with N digits of a 6 and ending in a 7, the square of that number will have $N + 1$ digits of 4, then N digits of 8, and the last digit will be a 9.

So why are these numbers important? In 1985, Donald E. Knuth held one of his regular problem-solving sessions at Stanford University. He asked if there were any numbers x and x^2 where the digits in both numbers were non-decreasing. Mathematicians sometimes ask these kinds of questions out of curiosity or to identify patterns. Occasionally, there are surprising results that have practical value, or the proof might involve a method of reasoning that is useful for solving other problems. So while you might find this question a bit strange, it is the kind of exercise that helps mathematically minded individuals sharpen their skills.

It was discovered numbers of the form 33...4 and 66...7 have the property, as it is easy to see the digits in the numbers are non-decreasing and the digits in the squares are also non-decreasing. Since there are infinitely many numbers of these forms, they could conclude there were infinitely many numbers for which x and x^2 have non-decreasing digits.

For more details, you can read the transcript of the 1985 session with Donald Knuth (search for "monotonic squares" or go to page 9) online: http://www-cs-faculty.stanford.edu/~uno/papers/cs1055.pdf

Practice Problems

$33{,}334^2$

$66{,}667^2$

Proof

Here is a proof for the pattern of 33...4 squared. A number starting with 3s and ending in a 4 is equal to $10^{n+1}/3 + 2/3$. Thus,

$$(33...4)^2 = (10^{n+1}/3 + 2/3)^2 = 10^{2n+2}/9 + 10^{n+1}(4/9) + 4/9$$

When we divide a power of 10 by 9, we get a number with only 1s plus a fraction of 1/9. So the first term has the digit 1 repeated $2n + 2$ times plus 1/9, and the middle term has the digit 4 repeated $n + 1$ times plus 4/9.

$$(33...4)^2 = (1...1 + 1/9) + (4...4 + 4/9) + 4/9$$

Now we add up the terms. There are $n + 1$ leading digits of 1; the next n digits are $5 = 1 + 4$; and the final digit is equal to 5 plus the fractions 1/9, 4/9, and 4/9, which neatly sum to 1. Thus the final digit is 6. So in the end, we have the result as desired.

$$(33...4)^2 = 1...14...45$$

Where there are $n + 1$ leading digits of 1, then n digits of 4, and a final digit of 5.

The proof for 66...7 is similar, using the fact that a number 66...7 can be written as $10^{n+1}(2/3) + 1/3$. So when we square the number, we have the following equation.

$(66...7)^2 = (10^{n+1}(2/3) + 1/3)^2 = 10^{2n+2}(4/9) + 10^{n+1}(4/9) + 1/9$

When we divide a power of 10 by 9, we get a number with only 1s plus a fraction of 1/9. So the first term has the digit 4 repeated $2n + 2$ times plus 4/9, and the middle term has the digit 4 repeated $n + 1$ times plus 4/9.

$(66...7)^2 = (4...4 + 4/9) + (4...4 + 4/9) + 1/9$

Now we add up the terms. There are $n + 1$ leading digits of 4; the next n digits are $8 = 4 + 4$; and the final digit is equal to 8 plus the fractions 4/9, 4/9, and 1/9, which neatly sum to 1. Thus the final digit is 9. So in the end, we have the result as desired.

$(33...4)^2 = 4...88...89$

Where there are $n + 1$ leading digits of 4, then n digits of 8, and a final digit of 9.

Solutions to Practice Problems

$33{,}334^2$. We need to write one more digit of 1 as there are digits of 3s, so the result starts out with five 1s. Then we write down as many 5s as there are 3s, so there are four 5s. And finally there is a 6. So the answer is 1,111,155,556.

$66{,}667^2$. We need to write one more digit of 4 as there are digits of 6s, so the result starts out with five 4s. Then we write down as many 8s as there are 6s, so there are four 6s. And finally there is a 9. So the answer is 4,444,488,889.

Part III: Multiplying Numbers

Multiply Two Numbers Ending In 5

What is 45 x 65? There is a shortcut to multiplication problems when both numbers end in the digit 5.

There are four steps. First, multiply the tens digits. Second, add the average of the tens digits. Third, multiply the result by 100. And finally add 25.

For 45 x 65, we first multiply the tens digits of 4 and 6 to get 24. Then we add the average of 5 to get 29. Multiplying by 100 results in 2,900, and finally adding 25 gets the result of 2,925.

Let's do another example of 35 x 75. We first multiply the tens digits of 3 and 7 to get 21. Then we add the average of 5 to get 26. Multiplying by 100 results in 2,600, and finally adding 25 gets the result of 2,625.

Let's do a final example of 55 x 85, we first multiply the tens digits of 5 and 8 to get 40. Then we add the average, which is $(5 + 8)/2 = 6.5$. It is okay that the average is not a whole number. Adding this to the product gets 46.5. Then we multiply by 100 to get 4,650, and finally adding 25 gets the result of 4,675.

The rule can also be extended to three-digit numbers or more. In that case, the rule is the same, with the "tens" digit being the number obtained by deleting the 5. For example, in the number 125, the "tens" digit would be the number 12.

The rule is sometimes written differently. If the numbers are $a5$ and $b5$, you can obtain the leading digits by adding the product ab to the average, $(a + b)/2$, rounded down. If a and b are both even or both odd, then the last two digits are 25. If one is one and one is even, the last two digits will be 75.

For example, let's do 55 x 85. The first part is the product $5 \times 8 = 40$ plus the average $(5 + 8)/2 = 6.5$, rounded down to 6. So we add 40 and 6 to get 46. Since 5 is odd and 8 is even, the last two digits will be 75. So the answer is $55 \times 85 = 4{,}675$. This is exactly the same result as before, but some people find this procedure easier to learn or quicker to use.

Practice Problems

15 x 35

85 x 65

35 x 45

95 x 5

105 x 145

Proof

Two numbers ending in 5 are written as $(10a + 5)$ and $(10b +5)$. When we multiply them, we have $(10a + 5)(10b + 5) = 100ab + 50(a + b) + 25$, which is equal to $100[ab + (a + b)/2] + 25$. That is, we compute the product plus the average as $ab + (a + b)/2$, multiply by 100, and then add 25.

Solutions to Practice Problems

15 x 35. The tens digits are 1 and 3. We add the product 1 x 3 = 3 and the average, 2, to get 5. Then we multiply by 100 to get 500 and add 25 to get 525.

85 x 65. The tens digits are 8 and 6. We add the product 8 x 6 = 48 and the average, 7, to get 55. Then we multiply by 100 to get 5,500 and add 25 to get 5,525.

35 x 45. The tens digits are 3 and 4. We add product of 3 x 4 = 12 and the average, 3.5, to get 15.5. Then we multiply by 100 to get 1,550 and add 25 to get 1,575.

95 x 5. We can write this problem as 95 x 05. The tens digits are 9 and 0. We add the product 9 x 0 = 0 and the average, 4.5, to get 4.5. Then we multiply by 100 to get 450 and add 25 to get 475.

105 x 145. The “tens” digits are found by deleting the 5 in each number to get the numbers 10 and 14. We add the product 10 x 14 = 140 and the average, 12, to get 152. Then we multiply by 100 to get 15,200 and add 25 to get 15,225.

Multiply Numbers Differing By 2

My friend was asked the following question in an investment bank interview: what is 14 times 16? The interviewer then put pressure on her to answer quickly. How was she supposed to do this without a calculator?

The trick was to see 14 x 16 is the same as $(15 - 1)(15 + 1)$, and the product of $(x + y)(x - y)$ is equal to $x^2 - y^2$. In other words, she was supposed to see that 14 x 16 is also equal to $15^2 - 1^2 = 225 - 1 = 224$.

We can use this trick for any numbers that differ by 2. The product of the two numbers will be the square of their average minus 1.

To do another example, 7 x 9 is equal to $8^2 - 1$. And in fact we can verify $7 \times 9 = 63 = 8^2 - 1 = 64 - 1$.

We can also combine this trick with the methods on how to square numbers quickly. For example, what is 51 x 53? Because 51 and 53 differ by 2, we know their product is equal to the square of their average of 52 minus 1. That is, 51 x 53 is equal to $52^2 - 1$. But what is 52 squared? We can use the trick of squaring numbers between 41 and 59. We add the difference of 2 to 25 to get 27, then we append the square of 2 written as a two-digit number, 04. So the square of 52 is 2,704, and therefore 51 x 53 = 2,704 – 1 = 2,703.

This rule can be extended to numbers that differ by 4, 6, or any even number. That is, the rule works whenever two numbers are equidistant from the same number. Their product will be the square of their average minus the square of their distance from the average. Note the distance to the average is half of the difference between the two numbers.

For example, what is 14 x 18? These numbers are equidistant from 16, with a distance of 2 from the average. Therefore, their product is equal to the square of 16 minus the square of 2. That is, 14 x 18 = $16^2 - 2^2$, which is equal to 256 – 4 = 252.

Similarly, we can calculate 13 x 21. The average of the two numbers is 17, and they are a distance of 4 from the average. Therefore, their product is equal to the square of 17 minus the square of 4. That is, we can compute 13 x 21 = $17^2 - 4^2$, which is equal to 289 – 16 = 273.

Practice Problems

11 x 13

18 x 16

13 x 19

98 x 102

46 x 54

Proof

Numbers at distance y from x are $(x - y)$ and $(x + y)$. The product of these numbers is $(x - y)(x + y) = x^2 - y^2$. That is, when two numbers are a distance y from their average x, their product is the square of the average minus the square of the distance from the average.

Solutions to Practice Problems

11 x 13. We re-write 11 x 13 = (12 – 1)(12 + 1), and then solve $12^2 - 1^2$, which is equal to 144 – 1 = 143.

18 x 16. We re-write 18 x 16 = (17 + 1)(17 – 1), and then solve $17^2 - 1^2$, which is equal to 289 – 1 = 288.

13 x 19. We re-write 13 x 19 = (16 – 3)(16 + 3), and then solve $16^2 - 3^2$, which is equal to 256 – 9 = 247.

98 x 102. We re-write 98 x 102 = (100 – 2)(100 + 2). Then we can solve $100^2 - 2^2$ = 10,000 – 4 = 9,996.

46 x 54. We re-write 46 x 54 = (50 – 4)(50 + 4), and then solve $50^2 - 4^2$, which is equal to 2,500 – 16 = 2,484.

Multiply Numbers Units Digits Sum To 10 And Same Tens Digit

In the last section, we solved 14 x 16 = $15^2 - 1^2$. This is a good way to solve the problem. But there is an alternate way that is perhaps even faster! This is one of my favorite tricks because the problems can be solved very easily.

You do, however, have to check the conditions for this trick. What you need is the units digits to sum to 10 and the tens digits to be the same. For example, in 14 x 16, the units digits of 4 and 6 do sum to 10, and the tens digit of 1 is the same for both numbers. This trick will when the units digits come from the pairs (1, 9), (2, 8), (3, 7), or (4, 6). It will also work for the pair (5, 5). In that case, both numbers will be the same, and the trick will be equivalent to squaring a number ending in 5.

While the trick has several conditions, these problems do arise, and there is a really fast and fun way to solve them.

Here is the procedure. You first multiply the tens digit by one more than itself. Then you multiply the units digits together and append the result, written as a two-digit number. And that's your answer!

For example, let's do 14 x 16. We multiply the tens digit of 1 by one more than itself, 2. So we have 1 x 2 = 2. Then we multiply the units digits of 4 and 6. This is 24, which we append to 2. So our answer is 224. That was easy!

Let's do another example. Let's try 33 x 37. We multiply the tens digit of 3 by one more than itself, 4. This is 3 x 4 = 12. Then we multiply the units digits of 3 and 7 to get 21, which we append to 12. Putting it together, we have 33 x 37 = 1,221. This is pretty cool!

You can also use the trick for three-digit or larger numbers. In that case, you need all the digits except the units digit to be the same (you can consider that the "tens" digit part). Then the procedure is the same: you multiply the "tens" digit part by one more than itself, and then you append the product of the units digit as the last two digits of the answer.

Let's do an example of 121 x 129. The “tens” digit in both numbers is the part not in the units digit. In both numbers the “tens” digit is 12. So to do the trick, we first multiply 12 by one more than itself, 13, to get 156. Now we need to multiply the units digits of 1 and 9, which we write as a two-digit number 09. Putting it together, we have 121 x 129 = 15,609.

The trick is very easy to do and impressive when you can pull it off, so keep an eye out for these types of problems.

(Incidentally, if you did not know how to do 12 x 13, there are tricks we have learned to solve this problem. One method is the trick to multiply numbers between 11 and 19. We add the units digit of 3 to the number 12 to get 15. This is multiplied by 10 to get 150. And then we would add the product of the units digits 2 x 3 = 6 to get the answer of 12 x 13 = 156. Another method is to split up the multiplication into parts by seeing that 13 = 12 + 1. Then 12 x 13 = 12 x (12 + 1) = 12^2 + 12 = 144 + 12 = 156.)

Practice Problems

12 x 18

21 x 29

43 x 47

98 x 92

56 x 54

Proof

Write one number as $10a + b$. The units digit of the other number should be $10 - b$ so that the units digits sum to 10. The tens digit should be the same, so the tens digit is $10a$. So the other number is $10a + (10 - b)$.

Multiplying the numbers $10a + b$ and $10a + (10 - b)$, we get the result of $(10a + b)[10a + (10 - b)] = 100a^2 + 10a(10 - b) + b(10a) + b(10 - b)$. We can cancel terms and group the remaining terms. Ultimately, this expression simplifies to $100a(a + 1) + b(10 - b)$. The term $a(a + 1)$ means to multiply the leading digit a by one more than itself, and that is multiplied by 100 so it starts in the hundreds spot. The term $b(10 - b)$ means to multiply the two units digits together, and that result becomes the last two digits of the answer.

Solutions to Practice Problems

12 x 18. We multiply 1 by one more than itself, 2, to get 2. Then we multiply the units digits, 2 x 8 = 16, to get the last two digits. The answer is 216.

21 x 29. We multiply 2 by one more than itself, 3, to get 6. Then we multiply the units digits, 1 x 9 = 9, to get the last two digits of 09. The answer is 609.

43 x 47. We multiply 4 by one more than itself, 5, to get 20. Then we multiply the units digits, 3 x 7 = 21, to get the last two digits. The answer is 2,021.

98 x 92. We multiply 9 by one more than itself, 10, to get 90. Then we multiply the units digits, so 8 x 2 = 16, to get the last two digits. The answer is 9,016.

56 x 54. We multiply 5 by one more than itself, 6, to get 30. Then we multiply the units digits, so 6 x 4 = 24, to get the last two digits. The answer is 3,024.

Multiply Numbers Tens Digits Sum To 10 And Same Units Digit

What is 34 x 74? There is a complement trick when the tens digits sum to 10 and the units digits are the same. For example, in 34 x 74, the units digits are both 4, while the tens digits of 3 and 7 sum to 10. This trick applies if the tens digits belong to the pairs (1, 9), (2, 8), (3, 7), or (4, 6). It will also work for the pair (5, 5), but in that case you will be squaring a number in the 50s, and it would be quicker to use that trick.

When the two numbers meet these conditions, you can solve the problem very quickly.

Here is the procedure. You multiply the tens digits together and then add the units digit. In 34 x 74, we calculate 3 x 7 = 21, and then we add 4 to get 25. These are the first two digits. Then we multiply the units digits (equivalent to squaring the units digit), and those are the last two digits. Returning to the example, $4 \times 4 = 4^2 = 16$. Appending 16 to 25 gets the answer 2,516.

Let's do another example of 46 x 66. We multiply the tens digits of 4 and 6 to get 24, to which the units digit of 6 is added to get 30. Then we square 6 to get 36. Appending 36 to 24 gets the answer 3,036.

Practice Problems

21 x 81

12 x 92

34 x 74

89 x 29

65 x 45

Proof

Write one number as $10a + b$. The units digit of the other number must be the same number b. The tens digit of the other number must sum to

10, so the tens digit should be $10 - a$. This means the other number is $10(10 - a) + b$.

Multiplying the numbers of $10 - a$ and $10(10 - a) + b$, we get the result of $(10a + b)[10(10 - a) + b] = 100a(10 - a) + 10ab + 10b(10 - a) + b^2$. We can cancel terms and group them to get $100[a(10 - a) + b] + b^2$. The term $a(10 - a) + b$ means to multiply the tens digits together and then add the units digit. That is multiplied by 100, which shifts the value over to the hundreds spot. The term b^2 means to multiply the units digits together, and those are the last two digits of the answer.

Solutions to Practice Problems

21 x 81. Multiply the tens digits together, 2 x 8 = 16, and then add the units digit of 1 to get 17 as the first two digits. Then square the units digit, $1^2 = 1$, and write that as the two-digit number 01. Appending 01 to 17 gets the result 1,701.

12 x 92. Multiply the tens digits together, 1 x 9 = 9, and then add the units digit of 2 to get 11 as the first two digits. Then square the units digit, $2^2 = 4$, and write that as the two-digit number 04. Appending 04 to 11 gets the result 1,104.

34 x 74. Multiply the tens digits together, 3 x 7 = 21, and then add the units digit of 4 to get 25 as the first two digits. Then square the units digit, $4^2 = 16$. Appending 16 to 25 gets the result 2,516.

89 x 29. Multiply the tens digits together, 8 x 2 = 16, and then add the units digit of 9 to get 25 as the first two digits. Then square the units digit, $9^2 = 81$, and write that as the two-digit number 81. Appending 85 to 25 gets the result 2,581.

65 x 45. Multiply the tens digits together, 6 x 4 = 24, and then add the units digit of 5 to get 29 as the first two digits. Then square the units digit, $5^2 = 25$. Appending 25 to 29 gets the result 2,925.

Multiply A Two-Digit Number By 11

What is 34 multiplied by 11? Here's a trick to solve the problem quickly.

There are three steps. First, copy the leading digit 3. Then, add up the two digits, 3 + 4 = 7. Finally, copy the second digit 4. Those numbers, written in order, are the digits of the answer. So 34 x 11 = 374.

To summarize, we can multiply a two-digit number by 11 by copying the first digit, adding the two digits together, and then copying the second digit. Those numbers written in order are the digits of the answer.

What about 57 times 11? We copy the first digit of 5. Then we have to add 5 and 7 which is 12. When the sum of the two digits is 10 or larger, we have to carry over the tens part to the previous digit of the answer. That is, from the sum of 12, we carry over the tens digit of 1 to make the previous digit of 5 into a 6. So we have 62 as the first two digits of the answer. Finally we copy the last digit of 7 to get 627.

Let's do another example. Let's multiply 76 times 11. We copy the first digit of 7. When we add 7 and 6 to get 13. Carrying over the tens digit of 1, the previous digit of 7 becomes an 8. So we have 83 as the first two digits of the answer. Finally we copy 6 as the last digit to get 836.

You can either carry over as you do the middle sum, or you can make the adjustment at the end. For example, we could do 38 x 11 by noting the answer is 3, 3 + 8, 8 = 3, 11, 8 and then carry over the 11 term to get the digits 4, 1, 8. Then we know 418 is the answer.

Practice Problems

21 x 11

45 x 11

81 x 11

56 x 11

97 x 11

Proof

A two-digit number can be written as $10a + b$. Multiplying by 11 is the same as multiplying by $(10 + 1)$.

This means $(10a + b)11 = (10a + b)(10 + 1) = 100a + 10b + 10a + b$, which is equal to $100a + 10(a + b) + b$. So we copy the digit a to the hundreds spot, then we take the sum $a + b$ as the tens spot, and finally we copy the digit b to the units spot.

Solutions to Practice Problems

21 x 11. We copy the digit 2, then we add 2 + 1 = 3, and then we copy the last digit 1. So the answer is 231.

45 x 11. We copy the digit 4, then we add 4 + 5 = 9, and then we copy the last digit 5. So the answer is 495.

81 x 11. We copy the digit 8, then we add 8 + 1 = 9, and then we copy the last digit 1. So the answer is 891.

56 x 11. We copy the digit 5. Then we add 5 + 6 = 11. The next digit is 1 and we carry over the 1 so the 5 becomes a 6. So our answer starts out as 61. Then we copy the last digit 6. So the result is 616.

97 x 11. We copy the digit 9. Then we add 9 + 7 = 16. The next digit is 6 and we carry over the 1 so the 9 becomes a 10. So our answer starts out as 106. Then we copy the last digit 7. So the result is 1,067.

Multiply Any Number By 11

What is 1,234 multiplied by 11? We can use a procedure similar to multiplying a two-digit number by 11.

The general process is: copy the first digit, add each pair of consecutive digits, and copy the last digit. If any sum is 10 or more, then carry over the digit in the tens spot to the previous digit of the answer.

Let's use this rule to multiply 1,234 by 11. We copy the first digit 1. Then we add each pair of consecutive digits. So we have 1 + 2 = 3, then we go to the next pair 2 + 3 = 5, and then 3 + 4 = 7. This means our middle digits are 357. Finally we copy the last digit of 4. Putting this all together, the answer is 13,574.

Let's do 567 times 11. We copy the first digit 5. Then we add each pair of consecutive digits. So we have 5 + 6 = 11, and 6 + 7 = 13. Then we copy the last digit of 7. So the answer has the digits 5, 11, 13, 7. Now we carry over for any digits that are 10 or larger. We carry over the 1 from the 11 so the digits are 6, 1, 13, 7. Then we carry over the 1 from the 13 so the digits are 6, 2, 3, 7. Now every digit is 9 or smaller, so these are the digits of the answer. Thus, 567 x 11 = 6,237.

As this example illustrates, it is a bit harder to use this method when the digits get larger and you need to carry over many times. It can be helpful to do the sums of consecutive numbers and then account for the carryover at the end.

Practice Problems

121 x 11

245 x 11

381 x 11

12,597 x 11

8,456 x 11

Proof

A number with k digits is written as $10^k a + 10^{k-1} b + 10^{k-2} c + \ldots + z$. Multiplying by 11 is the same as multiplying by 10 and then adding 1. Multiplying by 10 increases the power of 10 in each term. So the expression is $10^{k+1} a + 10^k b + 10^{k-1} c + \ldots + 10z$. Then we add back the original number. If we group the terms by the powers of 10, we end up with $10^{k+1} a + 10^k (a + b) + 10^{k-1}(b + c) + \ldots + 10(y + z) + z$. In other words, the highest power of 10 is copied from the first digit of the number a. Then each term is the sum of consecutive digits. And finally the last digit z is copied as the last digit of the answer.

Solutions to Practice Problems

121 x 11. We copy the digit 1. Then 1 + 2 = 3, and 2 + 1 = 3. Finally we copy 1 as the last digit. So the answer is 1,331.

245 x 11. We copy the digit 2. Then 2 + 4 = 6, and 4 + 5 = 9. Finally we copy 5 as the last digit. So the answer is 2,695.

381 x 11. We copy the digit 3. Then 3 + 8 = 11. We need to carry over the 1 so the previous digit 3 becomes a 4. So the answer starts out as 41. Then 8 + 1 = 9, and finally we copy 1 as the last digit. So the answer is 4,191.

12,597 x 11. This example also requires many carry-overs. So we'll do it without the carry-over and then make adjustments. We first copy 1, then we do the sums of 1 + 2 = 3, 2 + 5 = 7, 5 + 9 = 14, 9 + 7 = 16, and copy the last digit 7. So the results are 1, 3, 7, 14, 16, 7. We first carry over the 1 from the 14 to make 1, 3, 8, 4, 16, 7. Now we carry over the 1 from the 16 to make 1, 3, 8, 5, 6, 7. Now every digit is 9 or less, so the answer is 138,567.

8,456 x 11. This example requires many carry-overs. So we'll do it without the carry-over and then make adjustments. We first copy 8, then we do the sums of 8 + 4 = 12, 4 + 5 = 9, and 5 + 6 = 11, and copy the last digit 6. So the results are 8, 12, 9, 11, 6. We first carry over the 1 from the 12 to make 9, 2, 9, 11, 6. Now we carry over the 1 from the 11, so this increases the 9 into a 10, and we have 9, 2, 10, 1, 6. Now we carry over the 1 from the 10, so the 2 increases to a 3. This means 9, 3, 0, 1, 6. Now every digit is 9 or less, so the answer is 93,016.

Multiply Two Numbers In The 90s

Can you multiply 95 and 96 in your head? There is a wonderfully easy way to do this and similar problems.

There are three steps to the method. First, consider the difference of each number from 100. Second, subtract the difference of one number from the other number to get the first two digits of the answer. Finally, multiply the differences to get the last two digits of the answer.

For example, let's do 95 x 96. The first step is to consider the differences of each number from 100. The number 95 is 5 less than 100, and the number 96 is 4 less than 100. We then subtract the difference of one number from the other number. So we subtract 4 from 95 to get 91. We could also have subtracted 5 from 96 to get 91. Notice that both are the same so it doesn't matter which way we do it. This result of 91 becomes the first two digits of the answer. Finally, we multiply the differences. So we multiply 5 by 4 to get 20. These are the last two digits of the answer. Putting it together, we have 95 times 96 is equal to 9,120.

Let's do another example of 97 times 98. The first step is to consider the differences from 100. The number 97 is 3 less than 100 and the number 98 is 2 less than 100. Second, we subtract the difference of one number from the other number. So we have $95 = 97 - 2 = 98 - 3$. The final two digits can be found by multiplying the differences together. So we have 2 times 3 which equals 6. We have to remember to write this as a two-digit number, meaning the final two digits are 06. Putting the steps together, we have 97 times 98 is equal to 9,506.

Practice Problems

95 x 98

96 x 97

98 x 99

91 x 92

94^2

Proof

Numbers in the 90s can be written $100 - x$ and $100 - y$ for x and y less than 10. The product is $(100 - x)(100 - y) = 10{,}000 - 100y - 100x + xy$, which is equal to $100(100 - x - y) + xy$. The term $100 - x - y$ means to subtract the difference of one number from the other number. That term is multiplied by 100 so it becomes the first two digits. Then the product of the differences, xy, is the last two digits.

Solutions to Practice Problems

95 x 98. The number 95 is 5 less than 100 and the number 98 is 2 less than 100. So we subtract one difference from the other number to get that 95 – 2 = 93 = 98 – 5. So 93 are the first two digits. Then we take the product of the differences, 5 x 2 = 10. The result is 9,310.

96 x 97. The number 96 is 4 less than 100 and the number 97 is 3 less than 100. So we can do 96 – 3 = 93, or we can do 97 – 4 = 93. These are the first two digits. Then the last two digits are equal to the product of the differences, 4 x 3 = 12. The result is 9,312.

98 x 99. The number 98 is 2 less than 100 and the number 99 is 1 less than 100. So we can do 98 – 1 = 97, or we can do 99 – 2 = 97. These are the first two digits. Then the last two digits are 1 x 2 = 02 (remembering this product has to be written as a two-digit answer). The result is 9,702.

91 x 92. The number 91 is 9 less than 100 and the number 92 is 8 less than 100. So we can do 91 – 8 = 83, or we can do 92 – 9 = 83. These are the first two digits. Then the last two digits are 9 x 8 = 72. The result is 8,372.

94^2. Notice we can write this as 94 x 94. The number 94 is 6 less than 100. So we can do 94 – 6 = 88. These are the first two digits. Then the last two digits are 6 x 6 = 36. The result is 8,836. This problem illustrates the method can be used to square numbers in the 90s easily!

Multiply Two Numbers From 101 To 109

Can you multiply 102 and 107 in your head? There is a similar procedure to the one for multiplying numbers in the 90s. The only difference is the numbers are larger than 100, so we need to add the difference of one number to the other.

There are three steps to the method. First, consider the difference of each number from 100. Second, add the difference of one number to the other number. That result is the first three digits of the answer. Finally, multiply the differences, and that result becomes the last two digits of the answer.

For example, let's do 102 x 107. The number 102 is 2 more than 100, and the number 107 is 7 more than 100.

We add the difference of one number to the other number. So we add 2 to 107 to get 109. We could also add 7 to 102 to get 109. Notice that both results are the same so it doesn't matter which one we do. This result of 109 is the first part of our answer.

Finally, we multiply the differences. So we multiply 2 by 7 to get 14. These are the last two digits of the answer. Putting it together, we have 102 times 107 is equal to 10,914.

Take some time to learn this trick and the one about multiplying numbers in the 90s as the next few tricks will build upon the methods in these tricks.

Practice Problems

105 x 108

106 x 107

108 x 109

101 x 102

104^2

Proof

Two numbers in the 100s are $100 + x$ and $100 + y$ for x and y less than 10. The product is $(100 + x)(100 + y) = 10{,}000 + 100y + 100x + xy$, which is equal to $100(100 + x + y) + xy$. The term $100 + x + y$ is the part about adding one difference to the other number. That is multiplied by 100 so those numbers become the leading digits. Then the product of the differences, xy, is the last two digits.

Solutions to Practice Problems

105 x 108. The number 105 is 5 more than 100 and the number 108 is 8 more than 100. We add the difference of one number to the other number. So we can add 105 + 8 = 113, or we add 108 + 5 = 113. These are the first three digits. Then the last two digits are the product of the differences, 5 x 8 = 40. The result is 11,340.

106 x 107. The number 106 is 6 more than 100 and the number 107 is 7 more than 100. So we can do 106 + 7 = 113, or we can do 107 + 6 = 113. These are the first three digits. Then the last two digits are the product of the differences, 6 x 7 = 42. The result is 11,342.

108 x 109. The number 108 is 8 more than 100 and the number 109 is 9 more than 100. So we can do 108 + 9 = 117, or we can do 109 + 8 = 117. These are the first three digits. Then the last two digits are the product of the differences, 8 x 9 = 72. The result is 11,772.

101 x 102. The number 101 is 1 more than 100 and the number 102 is 2 more than 100. So we can do 101 + 2 = 103, or we can do 102 + 1 = 103. These are the first three digits. Then the last two digits are the product of the differences, 1 x 2 = 02 (remembering to write this as a two-digit number). The result is 10,302.

104^2. Notice we can write this as 104 x 104. The number 104 is 4 more than 100. So we can do 104 + 4 = 108. These are the first three digits. Then the last two digits are 4 x 4 = 16. The result is 10,816.

Multiply Two Numbers Near 1,000

Just as we generalized squaring numbers around 50 to the trick of squaring numbers around 500 and 5,000 and so on, we can generalize the trick for multiplying numbers around 100 to higher powers of 10.

The steps are the following. We consider the differences of the two numbers (to 1,000 or 10,000 or etc.). If the two numbers are smaller, we need to *subtract* the difference of one number from the other. If they are larger then we *add* the difference of one number to the other. Finally, we multiply the differences and write it with the appropriate number of digits (when the two numbers are smaller, we match the number of digits for the first part of our answer. When the two numbers are larger, we do one fewer digit than the first part of our answer).

For example, let's do 996 times 994. The differences from 1,000 are 4 and 6. Since the numbers are smaller, we will subtract 6 from 996 to get 990 as the first part of the answer. Then we multiply the differences to get 24, which we need to write as a three-digit number 024 (this is because we match the same number of digits in 990, which is a three-digit number). So the answer is 990,024.

Similarly, let's do 1,004 times 1,003. The differences from 1,000 are 4 and 3. We will add 3 to 1,004 to get 1,007 as the first part of the answer. Then we multiply the differences to get 12, which we need to write as a three-digit number 012 (this is because 1,007 has four digits and we need one fewer digit). So the answer is 1,007,012.

Practice Problems

995 x 998

1,006 x 1,007

998 x 999

1,001 x 1,002

$1{,}004^2$

Proof

Two numbers near 10^k, can be written as $10^k - x$ and $10^k - y$. Note the variables x and y can be positive, for numbers smaller than 10^k, or they can be negative, for numbers greater than 10^k. We multiply these numbers to get $(10^k - x)(10^k - y) = 10^{2k} - 10^k y - 10^k x + xy$, which is equal to $10^k(10^k - x - y) + xy$. The term $10^k - x - y$ is the part about compensating one difference to the other number. That is multiplied by 10^k so those numbers become the leading digits. Then the product of the differences, xy, is the last k digits.

Solutions to Practice Problems

995 x 998. The number 995 is 5 less than 1,000 and the number 998 is 2 less than 1,000. So we can do 995 – 2 = 993, or we can do 998 – 5 = 993. These are the leading digits. Then the last three digits are the product of the differences, 5 x 2 = 010 (remembering to write this as a three-digit number). The result is 993,010.

1,006 x 1,007. The number 1,006 is 6 more than 1,000 and the number 1,007 is 7 more than 1,000. So we can do 1,006 + 7 = 1,013, or we can do 1,007 + 6 = 1,013. These are the leading digits. Then the last three digits are the product of the differences, 6 x 7 = 042. The result is 1,013,042.

998 x 999. The number 998 is 2 less than 1,000 and the number 999 is 1 less than 1,000. So we can do 998 – 1 = 997, or we can do 999 – 2 = 997. These are the leading digits. Then the last three digits are the product of the differences, 2 x 1 = 002. The result is 997,002.

1,001 x 1,002. The number 1,001 is 1 more than 1,000 and the number 1,002 is 2 more than 1,000. So we can do 1,001 + 2 = 1,003, or we can do 1,002 + 1 = 1,003. These are the leading digits. Then the last three digits are the product of the differences, 1 x 2 = 002. The result is 1,003,002.

$1{,}004^2$. Notice we can write this as 1,004 x 1,004. The number 1,004 is 4 more than 1,000. So we can do 1,004 + 4 = 1,008. These are the leading digits. Then the last three digits are 4 x 4 = 016. The result is 1,008,016.

Multiply Two Numbers From 201 To 209, 301 To 309, Etc.

Can you multiply 202 and 207 in your head? We can modify the method to multiply numbers between 101 to 109, and this method will work for other numbers near multiples of 100 like numbers near 200, 300, etc.

There are four steps to the method. First, multiply the leading digits in both numbers. The numbers 202 and 207 have a leading digit of 2, so the first digit in the answer will be 2 x 2 = 4.

Second, consider the difference of each number from the closest multiple of 100. The number 202 is 2 more than 200, and the number 207 is 7 more than 200.

Third, add the differences together and multiply that by the leading digit. So we add 2 and 7 to get 9, and then multiply by the leading digit 2 to get 18. These are next two digits of the answer.

Fourth, multiply the differences of 2 and 7 to get 14. These are the last two digits.

Putting it all together, we have 202 x 207 = 41,814.

Let's do another example of 302 x 301. We multiply the leading digits to get 3 x 3 = 9.

Next, we consider the differences of 2 and 1, and then add them to get 3. We multiply by the leading digit of 3 to get 9. The result from this step is supposed to be a two-digit answer. So we write 9 as 09.

Finally, we multiply the differences to get 2 x 1 = 2. Again we write this as a two-digit answer so the final two digits are 02.

Putting it together, we have 302 x 301 = 90,902.

Let's do a third example which illustrates one more special case. Let's do 709 x 707. We multiply the leading digits to get 7 x 7 = 49. (Since the leading digits are always the same, this is the same as squaring the leading digit.)

Next, we add the differences of 9 and 7, which is 16. We multiply by 7 to get 112. The result from this step is supposed to be a two-digit answer. So we have to carry over the hundreds digit of 1 to the previous part of the answer. In other words, we add the value of the hundreds digit, 1, to the result from the last step, 49, to get 50. These are the leading digits of the answer. Then the two digits 12 are the next two digits of the answer. So the answer is something like 5,012 at this point.

Finally, we multiply the differences to get 9 x 7 = 63. These are the last two digits of the answer.

Putting it together, we have 709 x 707 = 501,263.

Here is the procedure in general to multiply numbers just above 200, 300, etc. Let Y denote the leading digit of the two numbers, so the two numbers are $Y0A$ and $Y0B$, for A and B between 1 and 9.

1. Multiply Y by itself, which is Y^2. The leading digit or digits of the answer will be the number Y^2.

2. Consider the difference of each number to the closest multiple of 100. This will be the numbers A and B.

3. Add the differences and multiply by Y. The next digits of the answer are $Y(A + B)$. Always write this as a two-digit number, so 6 would be written as 06. If the answer has three digits, then carry over the leading digit (in the hundreds spot) to the value of Y^2 from the previous step.

4. Finally, multiply the differences. The last two digits are A times B. Again, always write this as a two-digit number, so 6 would be written as 06.

Let's do a final example to illustrate the process. Let's do 909 x 908. The leading digit is $Y = 9$ and the differences are $A = 9$ and $B = 8$.

From the first step, we calculate $Y^2 = 81$.

Then we add the differences and multiply by the leading digit Y. So we have 9 x (9 + 8) = 9 x 17 = 153. This is a three-digit result, so we add the hundreds digit of 1 to the value of $Y^2 = 81$. Thus our answer starts out as 82 and then the next two digits are 53.

Finally, we multiply the differences to get 9 x 8 = 72. These are the last two digits.

Therefore, 909 x 908 = 825,372.

Practice Problems

205 x 208

306 x 307

408 x 409

601 x 602

704^2

Proof

Two numbers above a multiple of 100 can be written as $100Y + A$ and $100Y + B$. The product is equal to $(100Y + A)(100Y + B)$, which is $10{,}000Y^2 + 100YB + 100YA + AB = 10{,}000Y^2 + 100Y(A + B) + AB$. The term $10{,}000Y^2$ refers to squaring the leading digit. Then the next term $100Y(A + B)$ means to multiply A plus B by the leading digit Y, and then shift it over two places because it is multiplied by 100. Finally, the term AB is the product of the differences, which becomes the last two digits.

Solutions to Practice Problems

205 x 208. The number 205 has a difference of 5 from 200 and the number 208 has a difference of 8 from 200. First we square the leading digit of 2 to get $2^2 = 4$. Then we add the differences, 5 + 8 = 13, and multiply it by the leading digit 2, to get 13 x 2 = 26. Finally we multiply the differences, 5 x 8 = 40. Putting it all together, the answer is 42,640.

306 x 307. The number 306 has a difference of 6 from 300 and the number 307 has a difference of 7 from 300. First we square the leading digit of 3 to get $3^2 = 9$. Then we add the differences, 6 + 7 = 13, and multiply it by the leading digit 3, to get 13 x 3 = 39. Finally we multiply the differences, 6 x 7 = 42. Putting it all together, the answer is 93,942.

408 x 409. The number 408 has a difference of 8 from 400 and the

number 409 has a difference of 9 from 400. First we square the leading digit of 4 to get $4^2 = 16$. Then we add the differences, $8 + 9 = 17$, and multiply it by the leading digit 4, to get $17 \times 4 = 68$. Finally we multiply the differences, $8 \times 9 = 72$. Putting it all together, the answer is 166,872.

601 x 602. The number 601 has a difference of 1 from 600 and the number 602 has a difference of 2 from 600. First we square the leading digit of 6 to get $6^2 = 36$. Then we add the differences, $1 + 2 = 3$, and multiply it by the leading digit 6, to get $3 \times 6 = 18$. Finally we multiply the differences, $1 \times 2 = 02$ (remembering to write this as a two-digit number!). Putting it all together, the answer is 361,802.

704^2. This can be written as 704 x 704. The number 704 has a difference of 4 from 700. First we square the leading digit of 7 to get $7^2 = 49$. Then we add the differences, $4 + 4 = 8$, and multiply it by the leading digit 7, to get $8 \times 7 = 56$. Finally we multiply the differences, $4 \times 4 = 16$. Putting it all together, the answer is 495,616.

Multiply Two Numbers Just Below 200, 300, Etc.

Can you multiply 196 and 197? We can modify the method to multiply numbers in the 90s, and this method will work for other numbers near multiples of 100 like numbers near 200, 300, etc. It is not as easy to do mentally because you will have some hard intermediary multiplications to perform. With that warning, I have included the method for completeness.

There are three steps to the method. First, consider the difference of each number from the nearest multiple of 100, which is 200 in this example. The number 196 is 4 less than 200, and the number 197 is 3 less than 200.

Second, subtract the difference of one number from the other number. So we subtract 3 from 196 to get 193. We could also have subtracted 4 from 197 to get 193. Notice that both are the same so it doesn't matter which way we compensate the difference.

We then need to multiply this by the leading digit of the closest multiple of 100. Since the numbers are close to 200, the leading digit is 2. So we multiply 193 by 2 to get 386.

Finally, we multiply the differences. So we multiply 4 by 3 to get 12. These are the last two digits of the answer.

Putting it together, we have 196 times 197 equals 38,612.

Let's do another example of 299 times 298. First, consider the difference of each number from the nearest multiple of 100, which is 300. The number 299 is 1 less than 300 and the number 298 is 2 less.

Second, subtract the difference of one number from the other number. So we subtract 2 from 299 to get 297.

We then multiply this by the leading digit 3 from 300. So this results in $297 \times 3 = (300 - 3) \times 3 = 900 - 9 = 891$.

Finally, multiply the differences of 1 and 2. The result is 2, which needs to be written as a two-digit number, so we write it as 02.

Putting it all together, the result is 89,102.

So here are the steps to multiplying numbers just below $Y00$.

1. Calculate the differences from $Y00$, call them A and B.

2. Subtract the difference of one number from the other, which results in $Y00 - A - B$.

3. Multiply that result by Y. So we have $Y(Y00 - A - B)$ as the leading digits in the answer.

4. Multiply the differences of A and B, and write it as a two-digit result. These are the final two digits of the answer.

The step to calculate $Y(Y00 - A - B)$ is not always easy to do in your head. It can sometimes be made easier if you split up the multiplication, as will be illustrated in the solutions to the practice problems.

Practice Problems

195 x 198

296 x 297

398 x 399

591 x 592

694^2

Proof

Two numbers below a multiple of 100 can be written as written as the terms $100Y - A$ and $100Y - B$ for A and B between 1 and 9. The product is $(100Y - A)(100Y - B) = 10{,}000Y^2 - 100YB - 100YA + AB$, which is equal to $100Y(100Y - A - B) + AB$. The term $100Y - A - B$ is the difference of one number from the other. This is multiplied by the leading digit Y, and then shifted over two places because it is multiplied by 100. Then the product AB is added as the last two digits of the answer.

Solutions to Practice Problems

195 x 198. The number 195 has a difference of 5 from 200 and the number 198 has a difference of 2 from 200. So we can do 195 – 2 = 193 or we can do 198 – 5 = 193. This is multiplied by the digit 2 (from 200) to get 386. The product of the differences is 5 x 2 = 10. So the result is 38,610.

296 x 297. The number 296 has a difference of 4 from 300 and the number 297 has a difference of 3 from 300. So we can do 296 – 3 = 293 or we can do 297 – 4 = 293. This is multiplied by the digit 3 (from 300) to get 3(293) = 3(300 – 7) = 3(300) – 3(7) = 900 – 21 = 879. The product of the differences is 4 x 3 = 12. So the result is 87,912.

398 x 399. The number 398 has a difference of 2 from 400 and the number 399 has a difference of 1 from 400. So we can do 398 – 1 = 397 or we can do 399 – 2 = 397. This is multiplied by the digit 4 (from 400) to get 4(397) = 4(400 – 3) = 1,600 – 12 = 1,588. The product of the differences is 2 x 1 = 02 (once more, remembering to write this as a two-digit number!). So the result is 158,802.

591 x 592. The number 591 has a difference of 9 from 600 and the number 592 has a difference of 8 from 600. So we can do 591 – 8 = 583 or we can do 592 – 9 = 583. This is multiplied by the digit 6 (from 600) to get 6(583) = 6(600 – 17) = 3,600 – 102 = 3,498. The product of the differences is 9 x 8 = 72. So the result is 349,872.

694^2. The number 694 can be written as 694 x 694. The number 694 has a difference of 6 from 700. So we can do 694 – 6 = 688. This is multiplied by the digit 7 (from 700) to get 7(688) = 7(700 – 12), which is equal to 4,900 – 84 = 4,816. The product of the differences is 6 x 6 = 36. So the result is 481,636.

Multiply Two Numbers Near 100 (One Below And One Above)

What is 96 times 107? So far we have been dealing with multiplying when both numbers are below 100, or both numbers are above 100. How can we solve the problem when one number is below and another is above? The process is similar to the tricks we've already learned. Let's go through an example with the steps.

There are three steps to the method. First, consider the difference of each number from 100. The number 96 is 4 less than 100, and the number 107 is 7 above 100.

Second, compensate one number with the difference of the other, and then subtract 1. We either subtract from the larger number or add from the smaller number. So we can either take 107 and subtract 4 to get 103, or we can take 96 and add 7 to get 103. We then need to subtract 1 to get 102. These are the first three digits of the answer.

We then multiply the differences together. So we multiply 4 by 7 to get 28. Then we subtract this number from 100 to get 72. These are the last two digits of the answer.

Putting it together, we have 96 times 107 equals 10,272.

Let's do another example of 99 times 103. First, consider the difference of each number from 100, which is 1 for 99 and 3 for 103.

Second, compensate one number with the difference of the other and subtract 1. So we take $99 + 3 = 103 - 1 = 102$, which we then subtract 1 more to get 101 as the first three digits.

Third, multiply the differences of 1 and 3. The result is 3, which needs to be subtracted from 100 to get 97 as the last two digits.

Putting it all together, the result is 10,197.

To compensate one number with the other difference, you always end up with a number that's in between the two numbers you started with. So you always make the smaller number larger, or you make the larger number smaller. If you're outside of this range, you've made a mistake!

So here are the steps to multiplying numbers where one is below 100 and another is above 100.

1. Calculate the differences from 100, call them A and B.

2. Compensate each number with the difference of the other, and then subtract 1.

3. Multiply the differences together and then subtract that number from 100.

4. The result from step 2 is the leading digits of the answer, and the result from step 3 is the last two digits.

Practice Problems

95 x 108

93 x 106

98 x 109

91 x 102

94 x 104

Proof

Two numbers above and below 100 can be written $100 + A$ and $100 - B$ for A and B less than 10. The product is $(100 + A)(100 - B)$, which is equal to $10{,}000 - 100B + 100A - AB$. To get to our formula, we will do a trick and add the term $100 - 100$ to this formula. We can do this because we are adding $0 = 100 - 100$, and adding 0 does not change the sum.

So the formula is now $10{,}000 - 100B + 100A - AB + 100 - 100$, which is equal to $100(100 - B + A - 1) + 100 - AB$. The term $100 - B + A - 1$ is the part about compensating one number with the difference of the other, and then subtracting 1. This is multiplied by 100 so it is shifted over. Then the last two digits are $100 - AB$.

Solutions to Practice Problems

95 x 108. The number 95 is 5 less than 100, and the number 108 is 8 more than 100. We can do 95 + 8 = 103 or 108 – 5 = 103. Then we subtract 1 to get 102. The product of the differences is 5 x 8 = 40, which is subtracted from 100 to get 100 – 40 = 60. So the answer is 10,260.

93 x 106. The number 93 is 7 less than 100, and the number 106 is 6 more than 100. We can do 93 + 6 = 99 or 106 – 7 = 99. Then we subtract 1 to get 98. The product of the differences is 7 x 6 = 42, which is subtracted from 100 to get 100 – 42 = 58. So the answer is 9,858.

98 x 109. The number 98 is 2 less than 100, and the number 109 is 9 more than 100. We can do 98 + 9 = 107 or 109 – 2 = 107. Then we subtract 1 to get 106. The product of the differences is 2 x 9 = 18, which is subtracted from 100 to get 100 – 18 = 82. So the answer is 10,682.

91 x 102. The number 91 is 9 less than 100, and the number 102 is 2 more than 100. We can do 91 + 2 = 93 or 102 – 9 = 93. Then we subtract 1 to get 92. The product of the differences is 9 x 2 = 18, which is subtracted from 100 to get 100 – 18 = 82. So the answer is 9,282.

94 x 104. The number 94 is 6 less than 100, and the number 104 is 4 more than 100. We can do 94 + 4 = 98 or 104 – 6 = 98. Then we subtract 1 to get 97. The product of the differences is 6 x 4 = 24, which is subtracted from 100 to get 100 – 24 = 76. So the answer is 9,776.

Multiply Two Numbers Near 200, 300, Etc. (One Below And One Above)

What is 296 times 307? We can modify the procedure for doing the problem of 96 times 107 when the numbers are near another multiple of 100 like 200, 300, etc.

There are three steps to the method. First, consider the difference of each number from the nearest multiple of 100, which is 300 in this case. The number 296 is 4 less than 300, and the number 307 is 7 more than 300.

Second, compensate one number with the difference of the other, multiply that by the leading digit of the multiple of 100, and then subtract one more. So we can either take 307 minus 4 to get 303; or we can do 296 plus 7, to get 303. This number is multiplied by 3, since both numbers are near 300, to get 909. Then we subtract 1 to get 908.

Third, multiply the differences together and subtract that from 100. So we multiply 4 by 7 to get 28. Then we subtract this number from 100 to get 72. These are the last two digits of the answer.

Putting it together, we have 296 times 307 equals 90,872.

Let's do another example of 399 times 403. First, consider the difference of each number from 400, which is 1 for 399 and 3 for 403.

Second, compensate one number with the difference of the other, multiply that by the leading digit of the nearest multiple of 100, and then subtract 1. So we take 403 – 1 to get 402. This is multiplied by 4, since the numbers are near 400, to get 1,608. We then subtract 1 to get 1,607.

Third, multiply the differences of 1 and 3 and subtract from 100. The result is 3, which needs to be subtracted from 100 to get 97 as the last two digits.

Putting it all together, the result is 160,797.

The part to be careful about is that you need to compensate the numbers, multiply that result, and then, and only then, should you subtract 1. That is, you want to subtract 1 as the last part of the second step. If you subtract before you multiply, you will get the wrong answer.

So here are the steps to multiplying numbers where one is below $Y00$ and another is above $Y00$.

1. Calculate the differences from $Y00$, call them A and B.

2. Compensate each number with the difference of the other. Then multiply by Y. And finally reduce the result by subtracting 1.

3. Multiply the differences together and then subtract that number from 100.

4. The result from step 2 is the leading digits of the answer, and the result from step 3 is the last two digits.

Practice Problems

295 x 308

396 x 407

498 x 509

591 x 602

694 x 704

Proof

Two numbers above and below $100Y$ can be written as $100Y + A$ and $100Y - B$ for A and B less than 10. We multiply these numbers to get $(100Y + A)(100Y - B) = 10{,}000Y^2 - 100YB + 100YA - AB$. To get to our formula, we will do the same trick of adding the term $100 - 100$ to this formula. We can legally do this because we are adding $0 = 100 - 100$.

So the formula is now $10{,}000Y^2 - 100YB + 100YA - AB + 100 - 100$, which is equal to $100[Y(100Y - B + A) - 1] + 100 - AB$. Inside the brackets, the term $Y(100Y - B + A) - 1$ is the part about compensating one number with the difference of the other, multiplying by Y, and then subtracting 1. This is multiplied by 100 which makes the result shifted over to be the leading digits of the answer. Then the term 100 – *AB* gives us the last two digits of the answer, equal to 100 minus the product of the differences.

Solutions to Practice Problems

295 x 308. The number 295 is 5 less than 300, and the number 308 is 8 more than 300. We can do 295 + 8 = 303 or 308 – 5 = 303. This is multiplied by 3 (from 300) to get 909. Then we subtract 1 to get 908. The product of the differences is 5 x 8 = 40, which is subtracted from 100 to get 100 – 40 = 60. So the answer is 90,860.

396 x 407. The number 396 is 4 less than 400, and the number 407 is 7 more than 400. We can do 396 + 7 = 403 or 407 – 4 = 403. This is multiplied by 4 (from 400) to get 1,612. Then we subtract 1 to get 1,611. The product of the differences is 4 x 7 = 28, which is subtracted from 100 to get 100 – 28 = 72. So the answer is 161,172.

498 x 509. The number 498 is 2 less than 500, and the number 509 is 9 more than 500. We can do 498 + 9 = 507 or 509 – 2 = 507. This is multiplied by 5 (from 500) to get 2,535. Then we subtract 1 to get 2,534. The product of the differences is 2 x 9 = 18, which is subtracted from 100 to get 100 – 18 = 82. So the answer is 253,482.

591 x 602. The number 591 is 9 less than 100, and the number 602 is 2 more than 100. We do 591 + 2 = 593 or 602 – 9 = 593. This is multiplied by 6 (from 600) to get 6(593) = 6(600 – 7) = 3,600 – 42 = 3,558. Then we subtract 1 to get 3,557. The product of the differences is 9 x 2 = 18, which is subtracted from 100 to get 100 – 18 = 82. So the answer is 355,782.

694 x 704. The number 694 is 6 less than 700, and the number 704 is 4 more than 700. We can do 694 + 4 = 698 or 704 – 6 = 698. This is multiplied by 7 (from 700) to get 4,886. Then we subtract 1 to get 4,885. The product of the differences is 6 x 4 = 24, which is subtracted from 100 to get 100 – 24 = 76. So the answer is 488,576.

Part IV: Dividing Numbers

Divisibility Rules

Divisibility rules help you determine if one number can evenly divide another without having to compute the answer to the division problem.

These are rules you probably learned in school, and we present them because they occasionally come in handy. Here are some of the most common divisibility rules.

A number is divisible by 10 if the last digit is a 0.

A number is divisible by 5 if the last digit is a 0 or 5.

A number is divisible by 2 if the last digit is an even number (0, 2, 4, 6, or 8).

A number is divisible by 4 if the last two digits are divisible by 4.

A number is divisible by 8 if the last three digits are divisible by 8.

A number is divisible by 3 if the sum of all of its digits is divisible by 3.

A number is divisible by 9 if the sum of all of its digits is divisible by 9.

A number is divisible by 6 if it is divisible by 2 and 3.

A number is divisible by 11 if the *alternating* sum of its digits is divisible by 11. Start with the first digit on the left. The alternating sum means to subtract and then add digits alternately. So the second digit is subtracted, then the third digit is added, and so on.

For example, consider the number 91,234. Start with the digit on the left, 9. Then subtract the next digit 1, add the next digit 2, subtract the next digit 3, and add the last digit 4. The equation is $9 - 1 + 2 - 3 + 4 = 11$. The result 11 is divisible by 11, so we know 91,234 is divisible by 11.

You can also take the alternating sum of digits from right to left, if that is easier. Or you can add up all the digits in odd spots (the first, third, fifth, etc.), then add up the digits in the even spots (the second, fourth, sixth, etc.) and subtract one sum from the other to get an alternating sum.

Practice Problems

Determine if the numbers are divisible by 2, 3, 4, 5, 6, 8, 9, 10, or 11.

1,235

3,384

Proof

The proofs employ modulo arithmetic and can be skipped if desired.

We write a *k*-digit number as $N = n_0 10^0 + n_1 10^1 + \dots + n_k 10^k$. To say N is divisible by some number, that means N is congruent to 0 modulo (or "mod") that number.

For divisibility by 10, we reduce by modulo 10. Positive powers of 10 are all equal to 0 mod 10, so any term with a power of 10^1 or higher vanishes. Therefore, here is how N reduces modulo 10.

$N = n_0 10^0 + n_1 10^1 + \dots + n_k 10^k \pmod{10}$

$N = n_0 \pmod{10}$

Then N is congruent to 0 if and only if $n_0 = 0$, which means the last digit of N must be equal to 0.

For divisibility by 5, we reduce by modulo 5. Positive powers of 10 are all equal to 0 mod 5, so again any term with a power of 10^1 or higher vanishes. Therefore, here is how N reduces modulo 5.

$N = n_0 10^0 + n_1 10^1 + \dots + n_k 10^k \pmod{5}$

$N = n_0 \pmod{5}$

Then N is congruent to 0 if and only if $n_0 = 0$ or 5, which means the last digit of N must be equal to 0 or 5.

For divisibility by 2, we reduce by modulo 2. Positive powers of 10 are all equal to 0 mod 2, so again any term with a power of 10^1 or higher vanishes. Therefore, here is how N reduces modulo 2.

$N = n_0 10^0 + n_1 10^1 + \ldots + n_k 10^k \pmod 2$

$N = n_0 \pmod 2$

Then N is congruent to 0 if and only if n_0 is an even number, which means the last digit of N must be equal to 0, 2, 4, 6, or 8.

For divisibility by 4, we reduce by modulo 4. Since 100 is divisible by 4, every power 10^2 and higher is equal to 0 mod 4, so any term with a power of 10^2 or higher vanishes. Therefore, here is how N reduces modulo 4.

$N = n_0 10^0 + n_1 10^1 + \ldots + n_k 10^k \pmod 4$

$N = n_0 + n_1 10^1 \pmod 4$

Then N is congruent to 0 if and only if $n_0 + n_1 10^1$ is divisible by 4. The digits n_0 and n_1 are the last two digits of N, which means the last two digits of N must be divisible by 4.

For divisibility by 8, we reduce by modulo 8. Since 1,000 is divisible by 8, every power 10^3 and higher is equal to 0 mod 8, so any term with a power of 10^3 or higher vanishes. Therefore, here is how N reduces modulo 4.

$N = n_0 10^0 + n_1 10^1 + \ldots + n_k 10^k \pmod 8$

$N = n_0 + n_1 10^1 + n_2 10^2 \pmod 8$

Then N is congruent to 0 if and only if $n_0 + n_1 10^1 + n_2 10^2$ is divisible by 8. The digits n_0, n_1, and n_2 are the last three digits of N, which means the last three digits of N must be divisible by 8.

For divisibility by 3, we reduce by modulo 3. Note that every positive power of 10 is congruent to 1, because the terms 10, 100, etc. are all one more than 9, 99, etc., which are multiples of 3. Therefore,

$N = n_0 10^0 + n_1 10^1 + \ldots + n_k 10^k \pmod 3$

$N = n_0 + n_1 + \ldots + n_k \pmod 3$

So N is congruent to 0 if and only if the sum of its digits is divisible by 3.

A very similar proof shows the rule for divisibility by 9. Note the terms 10, 100, etc. are all one more than 9, 99, etc. which are multiples of 9. So when we reduce by modulo 9, we also get the sum of the digits of N, which needs to be divisible by 9 in order that the original number be divisible by 9.

A number is divisible by 6 if and only if it is divisible by 2 and 3. This follows because 6 = 2 x 3, so a number is divisible by 6 if and only if it is divisible by both its prime factors 2 and 3.

For divisibility by 11, we reduce by modulo 11. Consider the powers of 10. Obviously $10^0 = 1$ is congruent to 1 mod 11. Then $10^1 = 10$ is congruent to 10 mod 11 = -1 modulo 11. We can find 10^2 by noting that $10^2 = 10(10^1) \bmod 11 = 10(-1) \bmod 11 = -10 \bmod 11 = 1 \bmod 11$. By a similar process, we can find 10^3 is equal to -1 mod 11.

So there is a pattern. The even powers of 10, such as 10^0, 10^2, and so on, are all congruent to 1 modulo 11, and the odd powers 10^1, 10^3, and so on, are all congruent to -1 modulo 11. Therefore,

$$N = n_0 10^0 + n_1 10^1 + \ldots + n_k 10^k \pmod{11}$$

$$N = n_0 - n_1 + \ldots + (-1)^k n_k \pmod{11}$$

Then N is congruent to 0 if and only if the alternating sum of its digits, taken from right to left, is divisible by 11. Multiplying by -1 gives the same result when the alternating sum is taken from left to right.

Solutions to Practice Problems

1,235. The number is divisible only by 5. The last digit is 5 (so not divisible by 2, 4, 6, 8, or 10); and the sum of the digits is 11 (so not divisible by 3 or 9). Finally, the alternating sum is 1 – 2 + 3 – 5 = -3, which is not divisible by 11 (so not divisible by 11).

3,384. The number is divisible by 2, 3, 4, 6, 8, and 9. The last digit is 4 (divisible by 2 but not by 10); the last two digits are 84 (divisible by 4); the last three digits are 384 (divisible by 8); the sum of the digits is 18 (divisible by 3 and 9). Since it is divisible by 2 and 3, it is also divisible by 6. Finally, the alternating sum is 3 – 3 + 8 – 4 = 4, which is not divisible by 11 (so not divisible by 11).

Divide By 9

It is easy to divide a number between 1 and 8 by 9. The decimal is the same number repeated. For example, 4/9 is 0.444...

What about dividing larger numbers? What is 1,231 divided by 9? You can try to do this problem by long division. This section is about a trick to do this problem very quickly.

Here is how we will do the problem. To do 1231/9, the first digit in the answer is the same as the first digit of 1,231. So the first digit is 1. The next digit is found by adding this 1 to the second digit of 1,231. So we add 1 and 2 to get 3 as the next digit. We repeat this process: to get the next digit, we will add the 3 to the third digit of 1,231 which is 3. So we have 6. The whole part of our answer is 136. We then add the 6 to the 1, the final digit of 1,231 to get the remainder of 7. Therefore, we can conclude 1231/9 = 136, remainder 7. Recalling the remainder is a fractional part of 9, we can also write this as 136 + 7/9 = 136.777...

Here is how we do the trick. We copy the first digit of the dividend. Then we keep adding the previous digit of our answer to the next digit of the dividend. When we get to the final digit, that portion is a remainder.

Let's try another example. Let's do 350 divided by 9. The first digit of the answer is 3. The next digit is 8 = 3 + 5. Then we add 8 + 0 = 8 as the remainder. Thus, the answer is 38, remainder 8 = 38 + 8/9 = 38.888...

Occasionally, we will have a sum that exceeds 10. In that case, we adjust over the extra 1 to the previous digit and add one more to the units digit to get the next part of the answer.

Let's try an example of 782 divided by 9. The first digit is 7. Then we add 7 + 8 = 15. Since this number is 10 or more, we need to increase the previous digit in our answer of 7 by 1. We also increment the units digit of 5 by 1 to be 6 to get the next digit in the answer. So our answer starts out as 86. Finally we add 6 to 2 to get a remainder of 8. Thus the answer is 86, remainder 8 = 86 + 8/9 = 85.888...

A final caveat is if the remainder is 9 or more, then you should subtract out 9 and increase the previous digit. For example, let's do 19/9. We copy the 1 for the first digit. Then we have 1 + 9 = 10 as the remainder. So we have 1, remainder 10 = 1 + 10/9 = 1 + 9/9 + 1/9 = 2 + 1/9. In other words, we subtracted the remainder of 10 by 9 to get 1, and then we incremented the previous digit by 1 as well.

Let's try another problem. Let's do 20,457 divided by 9. The first digit we copy as 2. Then we have 0 + 2 = 2 as the next digit, and 2 + 4 = 6 as the next digit, and then 6 + 5 = 11. Because 11 is larger than 9, we adjust over the 1 to increment the previous digit of 6 to be 7. The current digit is the units digit incremented by 1, so that is 2. So our answer at this point is 2272. We add the 2 to the final digit of the dividend, 7, to get the remainder of 9. Since this remainder is 9, we have 9/9, which will add 1 to the previous digit. Therefore, our answer is 2,273.

Here is a summary of how to divide by 9 using this trick. When the dividend has N digits, the answer will have one less digit, or $N - 1$ digits. The first digit of the answer is the same as that of the dividend. For the next few steps, the next digit of the answer is found as the sum of the previous digit of the answer with the next digit of the dividend. This process continues until you get $N - 1$ digits. If you get a result of 10 or more, you need to adjust by incrementing the previous digit of the answer and retaining 1 more than the units digit as the current digit in the answer. The sum with the final digit of the dividend is the remainder. If the remainder is 9 or more, then you will have to increase the final digit of the answer (this is very much like the carry-over process).

Practice Problems

321 ÷ 9

2,500 ÷ 9

1,109 ÷ 9

1,234 ÷ 9

2,082 ÷ 9

Proof

Many of the proofs in the section on division tricks are more complicated and also require familiarity with infinite geometric series. None of the proofs are necessary for doing the tricks, so it is okay to skip the proofs.

Before I get to this proof, I want to give the intuition using one example.

Let's say we want to divide 123 by 9. The trick states the answer will be the first digit 1, then the sum $1 + 2 = 3$, with a remainder of $3 + 3 = 6$. In other words, we can evaluate 123/9 by copying the first digit, successively adding the next digit, with the last sum as the remainder. Why does this work?

Let's figure out where this process comes from. The first trick is that we will re-write the fraction 1/9 as the fraction $(1/10)/(1 - 1/10)$. While this looks more complicated, there is a reason we are writing it this way.

The second trick is to use the sum of a geometric series. If x is between 0 and 1, then $1 + x + x^2 + x^3 + \ldots = 1/(1 - x)$. For example, when $x = 1/10$, then $1 + (1/10) + (1/10)^2 + (1/10)^3 + \ldots = 1/(1 - 1/10)$. So we can re-write $(1/10)/(1 - 1/10) = 1/10 + (1/10)^2 + (1/10)^3 + \ldots$

So when we divide 123 by 9, that is the same as $123(1/10)/(1 - 1/10)$. We can then expand out the term $1/(1 - 1/10)$ using the formula for the sum of a geometric series.

So we have $(123/10)(1 + (1/10) + (1/10)^2 + (1/10)^3 + \ldots)$, and this is then equal to $123/10 + 123/10^2 + 123/10^3 + \ldots = 12.3 + 1.23 + 0.123 + \ldots$

In other words, we can divide 123 by 9 by adding up 123 divided by successive powers of 10. Dividing by 10 means to shift the decimal point over once, dividing by 100 means to shift it over two times, and so on. When we add up the terms 12.3, 1.23, 0.123, and so on, there is a pattern. The leading digit in the tens place 1—which is the same leading digit of 123—the next digit in the units place is the sum of 2 and 1—and then the first decimal point is $3 + 2 + 1 = 6$—which is the sum of the three digits in 123. When dividing by 9 the value in the first decimal point keeps repeating, so we can conclude every decimal point thereafter is 6 as well, equivalent to saying there is a remainder of 6.

So let's generalize this example into a proof. Let's say a number N is divided by 9. As explained above, dividing by 9 is the same as multiplying by the infinite series $1/10 + (1/10)^2 + (1/10)^3 + \ldots$. So dividing N by 9 is equal to $N/10 + N/10^2 + N/10^3 + \ldots$

Decimal numbers are written as $N = n_k 10^k + n_{k-1} 10^{k-1} + \ldots + n_0$. Substituting into the infinite series, the result is the following:

$$(n_k 10^k + n_{k-1} 10^{k-1} + \ldots + n_0)/10 + (n_k 10^k + n_{k-1} 10^{k-1} + \ldots + n_0)/10^2 + \ldots$$

When we divide through by powers of 10, we then get:

$$(n_k 10^{k-1} + n_{k-1} 10^{k-2} + \ldots + n_0/10) + (n_k 10^{k-2} + n_{k-1} 10^{k-3} + \ldots + n_0/10^2) + \ldots$$

Now we inspect the summation and group terms by powers of 10. The leading digit of the answer has the highest power of 10, namely 10^{k-1}, and the only term is $n_k 10^{k-1}$. This is interpreted as saying the leading digit of the answer is equal to the leading digit of the original number. The next digit has the next higher power of 10, namely 10^{k-2}, which is the sum of n_k and n_{k-1}, which is the sum of the first two digits of the number N. If you write out the infinite series in more terms, you will also see that each successive digit includes the sum of the next digit of N.

Once the whole part of the divisor is calculated, the remainder is the sum of all the digits of N, and that digit will repeat indefinitely in the decimal representation because that is the pattern when dividing by 9. So we can take the sum of the digits as a remainder.

The formula also explains the adjustment rule: when the sum of digits is 10 or more, then we need to add that to the higher power of 10 and adjust the result accordingly.

Solutions to Practice Problems

321 ÷ 9. We copy the first digit 3, then we add 3 + 2 = 5 for the next step. The remainder is 5 + 1 = 6. Thus, the answer is 35, remainder 6, or 35 + 6/9 = 35 + 2/3.

2,500 ÷ 9. We copy the first digit 2, then we add 2 + 5 = 7 for the next step, and 7 + 0 = 7 as the final whole part. The remainder is 7 + 0 = 7. Thus, the answer is 277, remainder 7, or 277 + 7/9.

1,109 ÷ 9. We copy the first digit 1, then we add 1 + 1 = 2 for the next step, and 2 + 0 = 2 for the final whole part. The remainder is 2 + 9 = 11. Note this is larger than 9, so we need to adjust. We add the 1 from the tens place to make the previous 2 a 3, and we increase the 1 from the units place for a remainder of 2. Thus, the answer is 123, remainder 2, or 123 + 2/9.

1,234 ÷ 9. We copy the first digit 1, then we add 1 + 2 = 3 for the next step, and 3 + 3 = 6 for the final whole part. The remainder is 6 + 4 = 10. Note this is larger than 9, so we need to adjust. We add the 1 from the tens place to make the previous 6 a 7, and we increase the 0 from the units place for a remainder of 1. Thus, the answer is 137, remainder 1, or 137 + 1/9.

2,082 ÷ 9. We copy the first digit 2, then we add 2 + 0 = 2 for the next step, and 2 + 8 = 10 for the final whole part. Note this is larger than 9, so we need to adjust. We add the 1 from the tens place to make the previous 2 a 3, and we increase the 0 from the units place to make 1 (this is the next digit of the answer). So we then add 1 + 2 = 3 for the remainder. Thus, the answer is 231, remainder 3, or 231 + 3/9 = 213 + 1/3.

Find The Decimal Part When Dividing By 7

Here are the decimal parts when dividing a number between 1 and 6 by 7.

1/7 = 0.142857142857...

2/7 = 0.285714285714...

3/7 = 0.428571428571...

4/7 = 0.571428571428...

5/7 = 0.714285714285...

6/7 = 0.857142857142...

There is a pattern that can help you remember the decimal parts. The key is to memorize the digits 142857. The fraction 1/7 has these digits repeating in that order. The next largest fraction, 2/7, starts with the second largest digit, includes the rest of the digits in that order, and then those digits repeat in that order. So 2/7 starts with the 2, so it starts 2857, and then the digits 142857 repeat after that. Then 3/7 starts with the third highest digit, so it starts 42857, and then continues 142857. The pattern continues: the next highest fraction starts from the place of the next highest digit, and then it continues 142857.

In other words, 1/7 starts with 1, 2/7 starts with 2, 3/7 starts with 4, 4/7 starts with 5, 5/7 starts with 7, and 6/7 starts with 8.

Find The Decimal Part When Dividing By 9, 99, Etc.

In the decimal system, it is very easy to divide a number by 10, 100, or so on. You move the decimal point of the number to the left a number of places equal to the number of 0s. For example, to do 156 divided by 100, we move the decimal point of 156 over to the left equal to the number of 0s in 100 which is 2. So 156 divided by 100 is 1.56, and we can similarly divide by any number starting with a 1 followed by 0s.

There is a math trick to divide by a number consisting of only the digit 9. You probably know the pattern to divide by 9: the digits of the dividend keep repeating. For example:

1/9 = 0.1111...

2/9 = 0.2222...

3/9 = 0.3333...

4/9 = 0.4444...

5/9 = 0.5555...

6/9 = 0.6666...

7/9 = 0.7777...

8/9 = 0.8888...

When you divide by 99, there is a similar pattern. Here is how we do the trick: we write the dividend as a two-digit number, and then those two digits keep repeating. That is xy/99 = 0.xyxyxy....

For example, 61/99 = 0.6161... and 73/99 = 0.7373... The only adjustment we have to remember is when we divide a single-digit number by 99. In that case, we need to write that number as a two-digit number with a leading 0. For example, let's do 5/99. We write 5 as a two-digit number 05, and then we calculate 5/99 = 0.0505... with the two digits 05 continuing to repeat.

When you divide by 999, which is a *three*-digit number, the pattern is to write the dividend as a *three*-digit number, and then those *three* digits keep repeating. That is xyz/999 = 0.xyzxyz.... When our dividend is less than three-digits, we need to write it as a three-digit number with the appropriate leading zeros. For instance, to do 17/999 we need to write the dividend as 017, and then we can calculate 17/999 = 0.017017... with the three digits 017 repeating.

We can generalize for dividing by a number 99..9, which is an *N*-digit number consisting only of 9s. The pattern is to write the dividend as an *N*-digit number, and then those *N* digits keep repeating. When our dividend is less than *N*-digits, we need to write it as an *N*-digit number with the appropriate leading zeros.

Practice Problems

31 ÷ 99

1,210 ÷ 9,999

1,324 ÷ 9,999

11,434 ÷ 99,999

22,082 ÷ 99,999

Proof

This proof is along the lines of the section about dividing by 9. It requires a bit more advanced mathematics and is okay to skip.

A number consisting of k digits of 9 is always one less than 10 raised to the power of k, that is, $10^k - 1$.

We will prove that $1/(10^k - 1)$ is equal to a decimal with the value 1 in every k^{th} spot, and every other spot is 0. That is $1/(10^k - 1) = 0.000...1...$ where there are $k - 1$ zeros followed by a 1, and this pattern repeats.

This can be proved using the formula for a geometric series. If x is in between 0 and 1, then $1 + x + x^2 + x^3 + \ldots = 1/(1 - x)$.

So let's re-write $1/(10^k - 1) = (1/10^k)/(1 - 1/10^k)$, and then use the formula for the sum of a geometric series to get $1/(10^k - 1)$ is equal to $(1/10^k)[1 + 1/10^k + 1/10^{2k} + \ldots] = 1/10^k + 1/10^{2k} + 1/10^{3k} + \ldots$

This infinite series is easy to evaluate. The term $1/10^k$ means the decimal has a value of 1 in the kth spot. The next term $1/10^{2k}$ means the decimal has a value of 1 in the $2k$ spot. The pattern is there is a 1 in the spots k, $2k$, $3k$, and every multiple of k, and the rest of the decimal values are just 0.

So we have just proved that $1/(10^k - 1) = 0.000...1...$ where there are k – 1 zeros followed by a 1, and this pattern repeats.

For any number N between 1 and less than $10^k - 1$, we can find the decimal value of N divided by $10^k - 1$ by multiplication. That is, we have $N/(10^k - 1) = N(0.000...1...)$. The pattern is easy to see: we will get the number N written as a k digit number that repeats indefinitely every k digits.

Solutions to Practice Problems

31 ÷ 99. The number 31 repeats in the decimal 0.313131...

1,210 ÷ 9,999. The number 1210 repeats in the decimal 0.12101210...

1,324 ÷ 9,999. The number 1324 repeats in the decimal 0.13241324...

11,434 ÷ 99,999. The number 11434 repeats in the decimal 0.1143411434...

22,082 ÷ 99,999. The number 22082 repeats in the decimal 0.2208222082...

Find The Decimal Part When Dividing By 11, 101, Etc.

The numbers 9, 99, etc. are special because if we add 1 to each number we get 10, 100, etc. which are the powers of 10. This was the reason we could divide by 9, 99, etc. in an easy pattern.

Similarly, the numbers 11, 101, etc. are special because if we subtract 1 to each number we get 10, 100, etc. which are the powers of 10. Accordingly, there is a trick to divide by 11, 101, etc. in an easy pattern.

Let's first examine dividing by 11.

1/11 = 0.090909...

2/11 = 0.181818...

3/11 = 0.272727...

4/11 = 0.363636...

5/11 = 0.454545...

6/11 = 0.545454...

7/11 = 0.636363...

8/11 = 0.727272...

9/11 = 0.818181...

10/11 = 0.909090...

When dividing by 11, there will be two digits that always repeat. One way to remember this is the two digits are the number multiplied by 9. For instance, to do 4/11, we remember 4 x 9 = 36, and therefore the two digits are 36 which repeat.

But there is another way to see the pattern, and this is useful because we can generalize the method to dividing by 101, 1001, etc.

Here is the pattern. A number x divided by 11 will have a decimal representation with two digits AB repeat, where $A = x - 1$ and $B = 9 - A$.

This might look complicated but it's easier to explain in words. When we divide by 11, we need to find a two-digit number AB that repeats. The first number A will be one less than x which is simple enough to calculate. The other digit B is a number that makes $A + B$ equal 9. We call the number B the 9s complement of A. Once you have the digit A, you can think about the number that brings A up to 9. The 9s complement can also be memorized as the pairs of numbers (0, 9), (1, 8), (2, 7), (3, 6), (4, 5).

So let's try this method to divide a number by 11. Let's do 8 divided by 11. We need to find a two-digit number. The first digit is one less so it will be 7. The next digit is the 9s complement, which is 2. Therefore, the two digits 72 will repeat, and 8/11 = 0.72...

This method may seem like more work, but it's useful because we can generalize it to dividing by 101, 1001, etc.

Let's now explain how to divide by 101. A number x divided by 101 will have a decimal representation with four digits $ABCD$ repeating, where $AB = x - 1$ and $C = 9 - A$ and $D = 9 - B$.

Again this might look complicated but it's easier to explain in words. When we divide by 101, we need to find a four digit number $ABCD$ that repeats. The first two digits AB will be one less than x which is simple enough to calculate (as usual we may need to add a leading zero to force AB to be a two-digit number. If x is 10, then $AB = 09$, for example). Then the digit C is the 9s complement of A, and the digit D is the 9s complement of B.

Let's do an example of 38/101. The answer will be a decimal with four digits repeating. The first two digits are 38 – 1 = 37. The next digit is the 9s complement of 3, which is 6. And the fourth digit is the 9s complement of 7, which is 2. Therefore, the digits 3762 repeat, and we have that 38/101 = 0.3762... The pattern is not that easily recognized so dividing by 101 can be quite the impressive trick!

Let us keep extending the trick. When you divide by 1,001, you are close to 1,000 which has *three* zeros. The number of repeating digits will be double that amount, so it will have *six* repeating digits. The pattern is to consider the dividend as a *three*-digit number (with leading zeros if necessary), and then subtract one from that number. Those will be the first three digits of the answer. The next three digits will be the 9s complement of each digit in order. Symbolically, $x/1001$ will be a decimal with six repeating digits $ABCDEF$, where $ABC = x - 1$ and the digit D is the 9s complement of A, the digit E is the 9s complement of B, and the digit F is the 9s complement of C.

We can generalize for dividing x by a number 10..1, which is 1 more than 10^N. The pattern is to write the dividend as an N-digit number, with leading zeros if necessary. The answer will be a repeating decimal with $2N$ digits repeating. The first N digits will be equal to x minus 1. The next N digits are the 9s complement of the first $N - 1$ digits in order.

Let's do an example of 127/10,001. We are dividing by 10,001 which is closest to 10,000 and has 4 zeros. So our answer will have double that, or 8 repeating digits, in total.

The first 4 digits will be one less than 127, which is 0126. Note we write a leading zero to make this part a four-digit number. The next 4 digits are the 9s complement of each digit in order. So we have 9873 because the 9s complements of 0, 1, 2, and 6 in order are 9, 8, 7, and 3.

So our 8 repeating digits are 01269873, and so 127/10,001 = 0. 01269873...

This is pretty neat! And since the pattern is not immediately obvious, it can make for an impressive mental math trick.

Practice Problems

31 ÷ 101

1,210 ÷ 10,001

1,324 ÷ 10,001

11,434 ÷ 100,001

22,082 ÷ 100,001

Proof

This proof is along the lines of dividing by a number with only 9s. Once again, it requires more advanced mathematics and is okay to skip.

A number 11, 101, and so on, is always one more than 10 raised to the k power, that is, $10^k + 1$.

We again use the sum for a geometric series $1 + x + x^2 + \ldots = 1/(1 - x)$. If we negate the value of x term by term, then we get the formula for an alternating geometric series, which is $1 - x + x^2 - x^3 + \ldots = 1/(1 + x)$.

So let's re-write $1/(10^k + 1) = (1/10^k)/(1 + 1/10^k)$, and then use the formula for the sum of an alternating geometric series for the denominator. We have $1/(10^k + 1) = (1/10^k)[1 - 1/10^k + 1/10^{2k} - \ldots]$, which is then equal to $1/10^k - 1/10^{2k} + 1/10^{3k} - 1/10^{4k}$ ...

This infinite series is where the pattern comes from. The term $1/10^k$ means the decimal has a value of 1 in the kth spot. Then we have to compensate with the next term $-1/10^{2k}$. Think about what happens when we subtract 1 from a number like 100. We end up with 99, which is to say, we reduce the 1 into a 0, and then the remaining terms are the 9s complement of 0. This is essentially what the term $-1/10^{2k}$ does in the decimal expression. We have to "borrow" the 1 from the previous decimal spot, and then we end up with the 9s complement of all the previous decimal values. Since the geometric series alternates signs, the same pattern holds with the terms $+1/10^{3k}$ and $-1/10^{4k}$.

For any number N between 1 and less than $10k - 1$, we can find the decimal value of N divided by $10^k - 1$ by multiplication. When we divide N by $(10^k + 1)$, that is equal to $N/10^k - N/10^{2k} + N/10^{3k} - N/10^{4k}$.... This is where the pattern in the rule comes from: the first k spots are the number N, but then the next k spots we have to subtract N. This works by "borrowing" from the previous number N (reducing it by 1), and then filling the remaining decimal values with the 9s complement of each digit in order. And then the pattern repeats every $2k$ decimal places.

(For example, let's illustrate with 13/101. The infinite alternating geometric series has the first two terms 13/100 – 13/10,000, which equals 0.13 – 0.0013 = 0.1287. So you can see the 0.12 is a result of having to reduce 0.13 when we subtract out 0.0013. The next digits are found by subtracting 13 from 100, which gets us the 9s complement of the number 12. Now every two terms thereafter are the same computation, except we are dividing by a larger power of 10. So we end up with the same digits 1287 repeating over and over.)

Solutions to Practice Problems

31 ÷ 101. The first two decimal spots are 31 – 1 = 30, and the remaining two are the 9s complement of each digit in order, which are 69. So the digits 3069 repeat and the result is 0.3069....

1,210 ÷ 10,001. The first four decimal spots are 1210 – 1 = 1209, and the remaining four are the 9s complement of each digit in order, which are 8790. So the digits 12098790 repeat and the result is 0.12098790....

1,324 ÷ 10,001. The first four decimal spots are 1324 – 1 = 1323, and the remaining four are the 9s complement of each digit in order, which are 8676. So the digits 13238676 repeat and the result is 0.13238676....

11,434 ÷ 100,001. The first five decimal spots are 11434 – 1 = 11433, and the remaining five are the 9s complement of each digit in order, which are 88566. So the digits 1143388566 repeat and the result is 0.1143388566....

22,082 ÷ 100,001. The first five decimal spots are 22082 – 1 = 22081, and the remaining five are the 9s complement of each digit in order, which are 77918. So the digits 2208177918 repeat and the result is 0.2208177918....

Find The Decimal Part When Dividing By 91

How are we going to divide a number by 91? Let's do a few examples to see if there is a pattern.

1/91 = 0.010989...

12/91 = 0.131868...

80/91 = 0.879120...

If you play around with a few more examples, here are some things you might notice. The decimal will always have 6 digits that repeat. Furthermore, the final 3 digits are the 9s complement of the first three digits in order. But how do we get those first three digits?

The pattern is not as easy to see. The somewhat surprising answer is we do the following steps. We multiply the dividend by 11 (a trick we already learned), write that as a three-digit number, and then subtract 1.

Equivalently the process to divide by 91 is to first multiply the dividend by 11, and then divide that result by 1,001 (another trick we already learned).

To refresh your memory, here's how to multiply by 11. For a number less than 10, you just write that number two times. So 8 x 11 = 88 for example. For a two-digit number, you will get a three-digit answer. You copy over the first digit, then you add the two digits, and then you copy the second digit. So to do 16 x 11, we copy the first digit 1, then we add 1 + 6 = 7 as the middle digit, and finally we copy the 6. The answer is then 16 x 11 = 176.

So here's how to divide by 91 written symbolically. When we divide x by 91, our answer will have 6 digits repeating $ABCDEF$. The first 3 digits are $ABC = 11x - 1$ and the final 3 digits are DEF which are the 9s complements of A, B, and C in order.

Let's try 13/91. We first multiply 13 by 11 to get 143. Then we subtract 1 to get 142. Then we take the 9s complement of each digit in order to get 857. Thus our answer is 13/91 = 0.142857... (In fact, because 91 = 13x7, we could have simplified 13/91 = 1/7 and solved the problem that way).

Let's try another example of 56/91. We first multiply 56 by 11 to get 616. (We copy over the 5, then the middle digit is 5 + 6 = 11. We carry over the 1 to the first digit 5 so the first digit of the answer is 6 and the middle digit is 1. For the final digit we copy the 6.) Then we subtract 1 to get 615. We then take the 9s complement of each digit in order to get 384. Thus our answer is 56/91 = 0.615384... As you can see, the hardest part in the process is multiplying by 11 when it involves carrying over.

For a final example, let's do 40/91. We multiply by 11 to get 440. Then we subtract 1 to get 439. Then we take the 9s complement in order which is 560. So our answer is 40/91 = 0.439560...

Practice Problems

31 ÷ 91

26 ÷ 91

4 ÷ 91

76 ÷ 91

89 ÷ 91

Proof

The trick works because 1,001 = 91 x 11. So to divide by 91 is the same as to multiply by 11 and then divide by 1,001. When we multiply by 11, we get a result which we treat as a three-digit number. Then we can divide by 1,001 by subtracting one from that number, and then finding the 9s complement of each of the digits in order. Those 6 digits in all repeat in the decimal expansion. This is the process described when dividing a number by 91.

Algebraically, 91 = 1,001/11, so dividing a number x by 91 means we have the expression $x/91 = x/(1001/11) = (11x)/1{,}001$. In other words, we can divide by 91 by multiplying by 11 and then dividing by 1,001, and we can use the tricks to multiply by 11 and divide by 1,001 that have already been explained.

Solutions to Practice Problems

31 ÷ 91. We multiply 31 by 11 to get 341. The first three digits of the decimal are 341 – 1 = 340, and the next three digits are the 9s complement of each digit in order, which is 659. So the digits 340659 repeat and the result is 0.340659...

26 ÷ 91. We multiply 26 by 11 to get 286. The first three digits of the decimal are 286 – 1 = 285, and the next three digits are the 9s complement of each digit in order, which is 714. So the digits 285714 repeat and the result is 0.285714.... Actually these digits are familiar! We could have recognized that 26/91 = (2 x 13)/(7 x 13) = 2/7, and we already know the decimal expansion when dividing by 7.

4 ÷ 91. We multiply 4 by 11 to get 44. We need to write this as a three-digit number 044. The first three digits of the decimal are 044 – 1 = 043, and the next three digits are the 9s complement of each digit in order, which is 956. So the digits 043956 repeat and the result is 0.043956...

76 ÷ 91. We multiply 76 by 11 to get 836 (we have to remember to carry over when the two digits sum to more than 9). The first three digits of the decimal are 836 – 1 = 835, and the next three digits are the 9s complement of each digit in order, which is 164. So the digits 835164 repeat and the result is 0.835164...

89 ÷ 91. We multiply 89 by 11 to get 979 (we have to remember to carry over when the two digits sum to more than 9). The first three digits of the decimal are 979 – 1 = 978, and the next three digits are the 9s complement of each digit in order, which is 021. So the digits 978021 repeat and the result is 0.978021...

Find The Decimal Part When Dividing By 19

What is 5 divided by 19? There is a method to solve this faster than the long division method.

The trick is we will repeatedly divide by 2, each time joining the remainder with the result to get the new value to divide by 2. This sounds really strange, but it works and it is very easy to calculate.

The first step is to divide 5 by 2. We will get 2 with a remainder of 1. The result of 2 is the first digit in the decimal expression of 5/19.

The next step is we join the remainder of 1 with the result of 2 to create the new value 12. We divide this number by 2 to get 6, with a remainder of 0. Then 6 is the next digit in the decimal expression of 5/19, so we have 5/19 = 0.26 at this point.

We repeat the process to find the next decimal point. That is, we join the previous remainder of 0 with the previous result of 6 to create the value 06. We divide this by 2 to get 3, again with a remainder of 0. Then 3 is the next digit in the decimal expression, so 5/19 = 0.263 at this point.

Here is the algorithm illustrated for a few more decimal places.

5/2 = 2, remainder 1 (new value is 12)

12/2 = 6, remainder 0 (new value is 06)

06/2 = 3, remainder 0 (new value is 03)

03/2 = 1, remainder 1 (new value is 11)

11/2 = 5, remainder 1 (new value is 15)

15/2 = 7, remainder 1 (new value is 17)

17/2 = 8, remainder 1 (new value is 18)

To find 5/19, we take the whole part from each division. So we can write 5/19 = 0.2631578... This is a surprising way to divide by 19, and all it requires is dividing by 2 at each step!

In fact, you can probably do method in your head. It is easy to divide a number by 2, and you know odd numbers will have a remainder of 1 and even numbers will have a remainder of 0. So you can mentally keep track of the new value to divide by and then continue to calculate the decimal values one by one.

For example, let's do 17/19. Here is how we would think about this in our head. First we divide 17 by 2, which is 8 with a remainder of 1. So we call out 0.8 as part of the decimal result. While we do that, we calculate the next value 18/2 = 9, remainder 0. So we say the next decimal is 9. We then calculate 09/2 = 4, remainder 1. So we say the next decimal is 4. With some practice this is possible to calculate the decimal values out as far as necessary until the pattern repeats. This can make for quite the impressive trick.

To recap, here are the steps to divide x by 19.

1. Divide x by 2. The whole part A is the first number of the decimal, and the remainder a should be noted.

2. Divide the number aA by 2 (which we mean the number where a is the first digit and A is the second digit). The whole part B is the next number in the decimal, and the remainder b should be noted.

3. Divide the number bB by 2. The whole part C is the next number of the decimal, and the remainder c should be noted.

4. Repeat the process. Then $x/19 = 0.ABCDEFGH....$ is the decimal expansion.

Practice Problems

$1 \div 19$

$13 \div 19$

Proof

The number 19 is one less than 20, which is why this trick works. We can write $19 = 20 - 1$.

Therefore, we have $1/19 = 1/(20 - 1)$. Now we multiply the numerator and denominator by $1/20$ get $1/(20 - 1) = (1/20)/(1 - 1/20)$.

We will use the formula for the sum of an infinite geometric series. If x is in between 0 and 1, then $1 + x + x^2 + x^3 + \ldots = 1/(1 - x)$.

So we have $(1/20)/(1 - 1/20) = (1/20)[1 + 1/20 + 1/20^2 + \ldots]$, which means $1/20 + 1/20^2 + 1/20^3 + \ldots = (1/2)/10 + (1/2)^2/10^2 + (1/2)^3/10^3 + \ldots$

To divide a number N by 19, therefore, we can express the division as an infinite series $N/19 = N(1/19) = (N/2)/10 + (N/2^2)/10^2 + (N/2^3)/10^3 + \ldots$

We can obtain the value for the first decimal point, corresponding to the term $1/10$, by dividing N by 2. The next decimal point, corresponding to the term $1/10^2$, is obtained by further halving $N/2$, and so on. If there is a remainder in any step, we add that to the following term in the next step. A remainder would be $(1/2^k)/(10)^k$, which needs to be added to the next term of $(1/2^{k+1})/(10)^{k+1}$. To make the denominators match, we multiply the remainder by 10. This is why our next new value is the remainder times 10 plus the previous whole part.

Solutions to Practice Problems

1 ÷ 19. We will write out the algorithm in detail.

1/2 = 0, remainder 1 (new value is 10)

10/2 = 5, remainder 0 (new value is 05)

05/2 = 2, remainder 1 (new value is 12)

12/2 = 6, remainder 0 (new value is 06)

06/2 = 3, remainder 0 (new value is 03)

03/2 = 1, remainder 1 (new value is 11)

11/2 = 5, remainder 1 (new value is 15)

15/2 = 7, remainder 1 (new value is 17)

17/2 = 8, remainder 1 (new value is 18)

18/2 = 9, remainder 0 (new value is 09)

09/2 = 4, remainder 1 (new value is 14)

14/2 = 7, remainder 0 (new value is 07)

07/2 = 3, remainder 1 (new value is 13)

13/2 = 6, remainder 1 (new value is 16)

16/2 = 8, remainder 0 (new value is 08)

08/2 = 4, remainder 0 (new value is 04)

04/2 = 2, remainder 0 (new value is 02)

02/2 = 1, remainder 0 (new value is 01)

The new value in this last step is 01, which is the same as the value divided by in the first step. So at this point the steps will repeat. To find 1/19, we take the whole part from each division. So we can conclude that 1/19 = 0.052631578947368421... and this repeats.

13 ÷ 19. Let's imagine we are doing this in our head. The remainder is 0 for even numbers and 1 for odd numbers. So we only need to focus on getting the whole parts, and then we can find the next new value by joining 0 or 1 to the front. Here is how we might think. We have 13/2 is 6, remainder 1. Then we have 16/2 is 8, remainder 0. Then 08/2 is 4, remainder 0. Then 04/2 is 2, remainder 0. So we can already calculate 13/19 = 0.6842, which is pretty impressive to calculate to 4 decimal places in your head. We can keep going to get the entire decimal expansion which repeats as 13/19 = 0.684210526315789473...

Find The Decimal Part When Dividing By 29, 39, Etc.

What is 5 divided by 29? We can modify the procedure of dividing by 19 to get a similar method when dividing by 29.

The trick is we will repeatedly divide by 3, each time joining the remainder with the whole part result to get the new value to divide by 3. Again this sounds strange, but it works and it is relatively easy to calculate.

The first step is to divide 5 by 3. We will get 1 with a remainder of 2. The result of 1 is the first digit in the decimal expression of 5/29.

The next step is we append the remainder of 2 with the result of 1 to create the number 21. We divide this number by 3 to get 7, with a remainder of 0. The value 7 is the next digit in the decimal expression of 5/29, so we have 5/29 = 0.17 at this point.

We repeat the process to find the next value. That is, we join the previous remainder of 0 with the previous result of 7 to create the number 07. We divide this number by 3 to get 2, with a remainder of 1. The value 2 is the next digit in the decimal expression of 5/29, so we have 5/29 = 0.172 at this point.

Here is the algorithm illustrated for a few more decimal places.

5/3 = 1, remainder 2 (new value is 21)

21/3 = 7, remainder 0 (new value is 07)

07/3 = 2, remainder 1 (new value is 12)

12/3 = 4, remainder 0 (new value is 04)

04/3 = 1, remainder 1 (new value is 11)

11/3 = 3, remainder 2 (new value is 23)

23/3 = 7, remainder 2 (new value is 27)

$27/3 = 9$, remainder 0 (new value is 09)

$09/3 = 3$, remainder 0 (new value is 03)

$03/3 = 1$, remainder 0 (new value is 01)

$01/3 = 0$, remainder 1 (new value is 10)

To find 5/29, we take the whole part from each division. So we have that the decimal expansion is $5/29 = 0.17241379310$... This is a remarkably easy way to divide by 29, and all it requires is dividing by 3 at each step!

Here are the steps to divide x by 29.

1. Divide x by 3. The whole part A is the first number of the decimal, and the remainder a should be noted.

2. Divide the number aA by 3 (which we mean the number where a is the first digit and A is the second digit). The whole part B is the second number of the decimal, and the remainder b should be noted.

3. Divide the number bB by 3. The whole part C is the third number of the decimal, and the remainder c should be noted.

4. Repeat the process. Then $x/29 = 0.ABCDEFGH$.... is the decimal expansion.

In fact, we can generalize the process to divide by 39, 49, 59, …, until 89. (We already learned how to divide by 99 in an easier method.)

To recap, here are the steps to divide x by $Y9$, where $Y = 1, 2, \ldots,$ or 8.

1. Divide x by $Y + 1$. The whole part A is the first number of the decimal, and the remainder a should be noted.

2. Divide the number aA by $Y + 1$ (which we mean the number where a is the first digit and A is the second digit). The whole part B is the next number of the decimal, and the remainder b should be noted.

3. Divide the number bB by $Y + 1$. The whole part C is the third number of the decimal, and the remainder c should be noted.

4. Repeat the process. Then $x/Y9 = 0.ABCDEFGH....$ is the decimal expansion.

Let's try an example of this by dividing 5 by 79. In this case we have that $Y = 7$, so we need to repeatedly divide by $8 = 7 + 1 = Y + 1$.

5/8 = 0, remainder 5 (new value is 50)

50/8 = 6, remainder 2 (new value is 26)

26/8 = 3, remainder 2 (new value is 23)

23/8 = 2, remainder 7 (new value is 72)

72/8 = 9, remainder 0 (new value is 09)

09/8 = 1, remainder 1 (new value is 11)

11/8 = 1, remainder 3 (new value is 31)

31/8 = 3, remainder 7 (new value is 73)

Now we take the whole parts of each step of division to get the result. So we have 5/79 = 0.06329113...

Practice Problems

Calculate the following to 4 decimal places.

$1 \div 29$

$2 \div 39$

$3 \div 49$

$14 \div 59$

$35 \div 79$

Proof

This trick is a generalization of dividing by 19.

A number $Y9$ is one less than $(Y+1)10$, which is why this trick works. We can write $Y9 = (Y+1)10 - 1$.

Therefore, we have $1/Y9 = 1/[(Y+1)10 - 1]$. Now we multiply the numerator and denominator by $1/[(Y+1)10]$. We get the expression that $1/[(Y+1)10 - 1] = (1/[(Y+1)10])/(1 - 1/[(Y+1)10])$.

We will use the formula for the sum of an infinite geometric series. If x is in between 0 and 1, then $1 + x + x^2 + x^3 + \ldots = 1/(1 - x)$.

So we have $(1/[(Y+1)10])/(1 - 1/[(Y+1)10])$ is equal to the infinite series $(1/[(Y+1)10])[1 + 1/[(Y+1)10] + 1/[(Y+1)10]^2 + \ldots]$.

This is equal to $1/[(Y+1)10] + 1/[(Y+1)10]^2 + 1/[(Y+1)10]^3 + \ldots$

Finally, we can re-write this as the following series.

$(1/(Y+1))/10 + (1/(Y+1)^2)/10^2 + (1/(Y+1)^3)/10^3 + \ldots$

To calculate $N/Y9$, therefore, we multiply the above expression by N to get the series $(N/(Y+1))/10 + (N/(Y+1)^2)/10^2 + (N/(Y+1)^3)/10^3 + \ldots$

We can obtain the value for the first decimal point, corresponding to the term 1/10, by dividing N by $Y + 1$. The next decimal point, corresponding to the term $1/10^2$, is obtained by further dividing by $Y + 1$, and so on. If there is a remainder in any step, we add that to the next step. A remainder r would be $(r/(Y + 1)^k)/(10)^k$, which needs to be added to the next term of $(1/(Y + 1)^{k+1})/(10)^{k+1}$. To make the denominators match, we multiply the remainder by 10. This is why our next new value is the remainder times 10 plus the previous whole part.

Solutions to Practice Problems

1 ÷ 29. We will write out the algorithm in detail. Because the denominator is 29, we need to divide by 2 + 1 = 3.

1/3 = 0, remainder 1 (new value is 10)

10/3 = 3, remainder 1 (new value is 13)

13/3 = 4, remainder 1 (new value is 14)

14/3 = 4, remainder 2 (new value is 24)

We could continue this process to calculate as many decimal places as we wish (until it repeats). To find 1/29, we take the whole part from each division. So we have 1/29 = 0.0344...

2 ÷ 39. Let's imagine we are doing this in our head. To divide by 39, we need to repeatedly divide by 4. We have 2/4 is 0, remainder 2. Then we have 20/4 is 5, remainder 0. Then 05/4 is 1, remainder 1. Then 11/4 is 2, remainder 3. So we have 2/39 = 0.0512...

3 ÷ 49. To divide by 49, we need to repeatedly divide by 5. We have 3/5 is 0, remainder 3. Then we have 30/5 is 6, remainder 0. Then 06/5 is 1, remainder 1. Then 11/5 is 2, remainder 1. So we have 3/49 = 0.0612...

14 ÷ 59. To divide by 59, we need to repeatedly divide by 6. We have 14/6 is 2, remainder 2. Then we have 22/6 is 3, remainder 4. Then 43/6 is 7, remainder 1. Then 17/6 is 2, remainder 5. So 14/59 = 0.2372...

35 ÷ 79. To divide by 79, we need to repeatedly divide by 8. We have 35/8 is 4, remainder 3. Then we have 34/8 is 4, remainder 2. Then 24/8 is 3, remainder 0. Then 03/8 is 0, remainder 3. So 35/79 = 0.4430...

Find The Decimal Part When Dividing By 21

What is 11 divided by 21? We explained how to divide by numbers 19, 29, 39, …, 89 by a fast method. These numbers were all 1 less than 20, 30, 40, …, 90, which are multiples of 10. There is a similar but slightly more complicated method for dividing by 21, 31, …, 81, which are one more than multiples of 10. (Note we explained a special trick for dividing by 91 in another section).

First, let's explain how to divide 11 by 21. The basic procedure is to divide by 2 repeatedly, getting the new value by multiplying the remainder by 10 and then subtracting the whole part. This is a bit more complicated than the procedure to divide by 19, so let's do an example to clarify what the method is.

The first step is to divide 11 by 2. We will get 5 with a remainder of 1. The result of 5 is the first digit in the decimal expression of 11/21.

The next step is to get the new value. This time we multiply the remainder of 1 by 10 and then subtract out the previous whole value of 5. So we have 10 – 5 = 5 as the new value. We divide the new value by 2 to get 2, with a remainder of 1. The value 2 is the next digit in the decimal expression of 11/19, so we have 11/19 = 0.52 at this point.

We repeat the process to find the next new value. That is, we multiply the remainder of 1 by 10 and then subtract out the previous whole value of 2. So we have 10 – 2 = 8 as the new value. We divide the new value by 2 to get 4, with a remainder of 0. We again have another tricky part in this process. When dividing by 21, we never want the remainder to be 0. We will instead say that 4, remainder 0 is equal to a whole value 3, remainder of 2. This is valid because 4 = 3 + 2/2, which is 3 with a remainder of 2. So we take the whole part to be 3, and the remainder to be 2. The decimal expansion of 11/21 = 0.523 and we continue the process. The next new value would be the remainder of 2 times 10 minus the previous whole value of 3. So the next new value would be 17.

To recap, the major differences are how to obtain the new value, and a procedure to never let the remainder be 0 in any of the steps.

Here is the algorithm illustrated for a few more decimal places.

11/2 = 5, remainder 1 (new value is 1 x 10 – 5 = 5)

5/2 = 2, remainder 1 (new value is 1 x 10 – 2 = 8)

8/2 = 4, remainder 0 = 3, remainder 2 (new value is 2 x 10 – 3 = 17)

17/2 = 8, remainder 1 (new value is 1 x 10 – 8 = 2)

2/2 = 0, remainder 2 (new value is 2 x 10 – 0 = 20)

20/2 = 9, remainder 2 (new value is 2 x 10 – 9 = 11)

11/2 = 5, remainder 1 (new value is 1 x 10 – 5 = 5)

To find 11/21, we take the whole part from each division. So we have 11/21 = 0.5238095...

The process is similar to dividing by 19, but there are a few differences. When dividing by 19, we would take A, remainder a and get the new value as aA, which was found by adding the whole value to 10 times more than the remainder. In other words, we were calculating $10a + A$. When dividing by 21, we find the new value by multiplying the remainder by 10 and then subtracting the whole part. That is, we calculate $10a - A$ as the new value.

The other difference is we never let the remainder in any step be equal to 0. If necessary, we always reduce the whole part and bump up the remainder.

To recap, here are the steps to divide x by 21.

1. Divide x by 2. The whole part A is the first number of the decimal, and the remainder a should be noted. In any step, we never let the remainder be 0. When needed, we then reduce the whole part by 1 and increase the remainder to be 2.

2. Divide the number $(10a - A)$ by 2. The whole part B is the second number of the decimal, and the remainder b should be noted. (Again, we never let the remainder be 0).

3. Divide the number $(10b - B)$ by 2. The whole part C is the third number of the decimal, and the remainder c should be noted.

4. Repeat the process. Then $x/21 = 0.ABCDEFGH....$ is the decimal expansion.

Practice Problems

Calculate the following to 4 decimal places.

$1 \div 21$

$17 \div 21$

Proof

Note that $21 = 20 + 1$. We will again use an infinite geometric series to derive the algorithm.

We have $1/21 = 1/(20 + 1)$. Now we multiply the numerator and denominator by $1/20$ get $1/(20 + 1) = (1/20)/(1 + 1/20)$.

We will use the formula for the sum of an infinite alternating geometric series, which is $1 - x + x^2 - x^3 + \ldots = 1/(1 + x)$.

So we have $(1/20)/(1 + 1/20) = (1/20)[1 - 1/20 + 1/20^2 - \ldots]$, which is equal to $1/20 - 1/20^2 + 1/20^3 - \ldots = (1/2)/10 - (1/2)^2/10^2 + (1/2)^3/10^3 - \ldots$

To divide a number N by 21, therefore, we can express the division as the infinite series $N/21 = N(1/21) = (N/2)/10 - (N/2^2)/10^2 + (N/2^3)/10^3 - \ldots$

We can obtain the value for the first decimal point, corresponding to the term $1/10$, by dividing N by 2. The next decimal point, corresponding to the term $1/10^2$, is obtained by further halving $N/2$, and subtracting that from the previous remainder times 10. This is for the following reason. For the remainder in any step (which we force if necessary), we add that to the next term. A remainder would be $(1/2^k)/(10)^k$, which needs to be added to the next term of $(1/2^{k+1})/(10)^{k+1}$. To make the denominators match, we multiply the remainder by 10. Since consecutive terms are always of opposite signs, we need to subtract the whole part from the remainder. This is why our next new value is the remainder times 10 minus the previous whole part.

We also force a remainder because we always need the next new value to be positive so there is a number from which we can subtract the previous whole part.

Solutions to Practice Problems

1 ÷ 21. Here are the steps written out in detail. In the second and third steps we have to avoid letting the remainder be 0.

1/2 = 0, remainder 1 (new value is 1 x 10 – 0 = 10)

10/2 = 4, remainder 2 (new value is 2 x 10 – 4 = 16)

16/2 = 7, remainder 2 (new value is 2 x 10 – 7 = 13)

13/2 = 6, remainder 1

To find 1/21, we take the whole part from each division. So we have that 1/21 = 0.0476...

17 ÷ 21. This procedure is not as easy to do in your head, so we'll write it out again.

17/2 = 8, remainder 1 (new value is 1 x 10 – 8 = 2)

2/2 = 0, remainder 2 (new value is 2 x 10 – 0 = 20)

20/2 = 9, remainder 2 (new value is 2 x 10 – 9 = 11)

11/2 = 5, remainder 1

To find 17/21, we take the whole part from each division. So we have that 17/21 = 0.8095...

Find The Decimal Part When Dividing By 31, 41, Etc.

What is 5 divided by 31? We can modify the procedure of dividing by 21 to find the result when dividing by 21. The only difference is we now repeatedly divide by 3. Again, we never let the remainder be equal to 0. If needed, we always decrease the whole part by 1 and then increase the remainder to be 3.

Let's divide 5 by 31 using the algorithm.

5/3 = 1, remainder 2 (new value is 2 x 10 – 1 = 19)

19/3 = 6, remainder 1 (new value is 1 x 10 – 6 = 4)

4/3 = 1, remainder 1 (new value is 1 x 10 – 1 = 9)

9/3 = 3, remainder 0 = 2, remainder 3 (new value is 3 x 10 – 2 = 28). Notice how we made the adjustment so the remainder was not 0.

28/3 = 9, remainder 1 (new value is 1 x 10 – 9 = 1)

To find 11/31, we take the whole part from each division. So we have 11/21 = 0.16129...

The process is similar to dividing by 21, except we are dividing by 3 at each step, and adjusting the remainder to be 3 instead of 0 at every step.

We can generalize to divide x by $Y1$ where $Y = 2, 3, \ldots,$ or 8.

1. Divide x by Y. The whole part A is the first number of the decimal, and the remainder a should be noted. In any step, we never let the remainder be 0. When needed, we then reduce the whole part by 1 and increase the remainder to be Y.

2. Divide the number $(10a – A)$ by Y. The whole part B is the second number of the decimal, and the remainder b should be noted. (Again, we never let the remainder be 0, so we adjust if needed).

3. Divide the number $(10b – B)$ by Y. The whole part C is the third

number of the decimal, and the remainder c should be noted.

4. Repeat the process. Then $x/Y1 = 0.ABCDEFGH....$ is the decimal expansion.

Practice Problems

Calculate the following to 4 decimal places.

$1 \div 31$

$2 \div 41$

$3 \div 51$

$14 \div 61$

$35 \div 81$

Proof

This is analogous to the proof for dividing by 21. The number $Y1$ is one more than $Y0$. We can write $Y1 = Y0 + 1$.

Thus $1/Y1 = 1/(Y0 + 1)$. Now we multiply the numerator and denominator by $1/Y0$ get $1/(Y0 + 1) = (1/Y0)/(1 + 1/Y0)$.

We will use the formula for the sum of an infinite alternating geometric series, which is $1 - x + x^2 - x^3 + \ldots = 1/(1 + x)$.

So we have $(1/Y0)/(1 + 1/Y0) = (1/Y0)[1 - 1/Y0 + 1/Y0^2 - \ldots]$, which is equal to $1/Y0 - 1/Y0^2 + 1/Y0^3 - \ldots$. We can re-write this as the infinite series $(1/Y)/10 - (1/Y)^2/10^2 + (1/Y)^3/10^3 - \ldots$

To divide a number N by $Y1$, we can calculate the infinite alternating geometric series $N/Y1 = N(1/Y1) = (N/Y)/10 - (N/Y^2)/10^2 + (N/Y^3)/10^3 - \ldots$

We can obtain the value for the first decimal point, corresponding to the term $1/10$, by dividing N by Y. The next decimal point, corresponding to the term $1/10^2$, is obtained by further dividing by Y, and subtracting that from the previous remainder times 10. This is for the following reason. For the remainder in any step (which we force if necessary), we add that to the next term. A remainder would be $(1/Y^k)/(10)^k$, which needs to be added to the next term of $(1/Y^{k+1})/(10)^{k+1}$. To make the denominators match, we multiply the remainder by 10. Since consecutive terms are always of opposite signs, we need to subtract the whole part from the remainder. This is why our next new value is the remainder times 10 minus the previous whole part.

Solutions to Practice Problems

1 ÷ 31. To divide by 31, we need to repeatedly divide by 3. Here are the steps written out in detail.

1/3 = 0, remainder 1 (new value is 1 x 10 – 0 = 10)

10/3 = 3, remainder 1 (new value is 1 x 10 – 3 = 7)

7/3 = 2, remainder 1 (new value is 1 x 10 – 2 = 8)

8/3 = 2, remainder 2

To find 1/31, we take the whole part from each division. So we have that 1/31 = 0.0322...

2 ÷ 41. To divide by 41, we need to repeatedly divide by 4. We write out the steps as follows, making sure to avoid letting any of the remainders be equal to 0.

2/4 = 0, remainder 2 (new value is 2 x 10 – 0 = 20)

20/4 = 4, remainder 4 (new value is 4 x 10 – 4 = 36)

36/4 = 8, remainder 4 (new value is 4 x 10 – 8 = 32)

32/4 = 7, remainder 4

To find 2/41, we take the whole part from each division. So we have that 4/41 = 0.0487...

3 ÷ 51. To divide by 51, we need to repeatedly divide by 5. We write out the steps as follows, making sure to avoid letting any of the remainders be equal to 0.

3/5 = 0, remainder 3 (new value is 3 x 10 – 0 = 30)

30/5 = 5, remainder 5 (new value is 5 x 10 – 5 = 45)

45/5 = 8, remainder 5 (new value is 5 x 10 – 8 = 42)

42/5 = 8, remainder 2

To find 3/51, we take the whole part from each division. So we have that 3/51 = 0.0588...

14 ÷ 61. To divide by 61, we need to repeatedly divide by 6. We write out the steps as follows, making sure to avoid letting any of the remainders be equal to 0.

14/6 = 2, remainder 2 (new value is 2 x 10 – 2 = 18)

18/6 = 2, remainder 6 (new value is 6 x 10 – 2 = 58)

58/6 = 9, remainder 4 (new value is 4 x 10 – 9 = 31)

31/6 = 5, remainder 1

To find 14/61, we take the whole part from each division. So we have that 14/61 = 0.2295...

35 ÷ 81. To divide by 81, we need to repeatedly divide by 8. We write out the steps as follows, making sure to avoid letting any of the remainders be equal to 0.

35/8 = 4, remainder 3 (new value is 3 x 10 – 4 = 26)

26/8 = 3, remainder 2 (new value is 2 x 10 – 3 = 17)

17/8 = 2, remainder 1 (new value is 1 x 10 – 2 = 8)

8/8 = 0, remainder 8

To find 35/81, we take the whole part from each division. So we have that 4/41 = 0.4320...

Part V: Concluding Tricks

Sum Odd Numbers

What is 1 + 3 + 5 + 7 + 9 + 11 + 13? It would not be too difficult to add up the numbers. But there is a trick that makes it even faster. The sum of the first n odd numbers is equal to n^2. For example,

$1 + 3 = 4 = 2^2$

$1 + 3 + 5 = 9 = 3^2$

$1 + 3 + 5 + 7 = 16 = 4^2$

$1 + 3 + 5 + 7 + 9 = 25 = 5^2$

So to return to the problem, we could count there are 7 numbers, which means $1 + 3 + 5 + 7 + 9 + 11 + 13 = 7^2 = 49$.

Practice Problems

Sum the odd numbers up to 101.

Sum the odd numbers up to 1,017.

Proof

The formula can be proven inductively. We have already verified it is true for several base cases. Assuming it is true for the first k odd numbers (up to the number $2k - 1$), we wish to prove it is true for the next odd number of $2k + 1$.

That is, we want to evaluate the sum $1 + 3 + 5 + \ldots + (2k - 1) + (2k + 1)$.

By the induction hypothesis, we know the sum up to $2k - 1$ is equal to k^2. So we have the following.

$1 + 3 + 5 + \ldots + (2k - 1) + (2k + 1) = k^2 + (2k + 1) = (k + 1)^2$

The last step is factoring $k^2 + 2k + 1$, so indeed adding the next odd number is equal to the square of the next number.

Solutions to Practice Problems

Sum the odd numbers up to 101. There are 100/2 = 50 odd numbers up to 100, so there are 51 odd numbers up to 101. The sum is $51^2 = 2,601$ (use the trick to square a number in the 50s!).

Sum the odd numbers up to 1,017.There are 1016/2 = 508 odd numbers up to 1016, so there are 509 odd numbers up to 1,017. The sum is then $509^2 = 259,081$ (use the trick to square a number near the 500s!).

Sum Even Numbers

What is $2 + 4 + 6 + 8 + 10 + 12 + 14$? You could probably add the numbers without much trouble. But there is a formula that is useful. The sum of the first n even numbers is equal to $n(n + 1)$. For example,

$2 + 4 = 6 = 2(3)$

$2 + 4 + 6 = 12 = 3(4)$

$2 + 4 + 6 + 8 = 20 = 4(5)$

So to return to the problem, we could count there are 7 numbers, which means $2 + 4 + 6 + 8 + 10 + 12 + 14 = 7(8) = 56$.

Practice Problems

Sum the even numbers up to 100.

Sum the even numbers up to 1,016.

Proof

The formula can be proven inductively. We have already verified it is true for several base cases. Assuming it is true for the first k even numbers (up to the number $2k$), we wish to prove it is true for the next even number of $2k + 2$.

That is, we want to evaluate the sum $2 + 4 + 6 + \ldots + 2k + (2k + 2)$.

By the induction hypothesis, we know the sum to $2k$ is equal to $k(k + 1)$. So we have the following.

$2 + 4 + 6 + \ldots + 2k + (2k + 2) = k(k + 1) + (2k + 2)$

$= k^2 + 3k + 2 = (k + 1)(k + 2)$

The last step is factoring $k^2 + 3k + 2$, so indeed the formula is verified for adding the next even number.

Solutions to Practice Problems

Sum the even numbers up to 100. There are 100/2 = 50 even numbers up to 100. The sum is $50(51) = 50^2 + 50 = 2,550$.

Sum the even numbers up to 1,016.There are 1016/2 = 508 even numbers up to 1016. The sum is $508(509) = 508^2 + 508$, which is then equal to 258,064 + 508 = 258,572 (use the trick for squaring a number near the 500s!).

Sum Consecutive Numbers

What is the sum of the whole numbers from 1 to 100? Just as we could figure out a pattern for the sum of odd and sum of even numbers, there is a pattern for the sum of all whole numbers.

The trick is to pair the numbers 1 to 100 as (1, 100), (2, 99), (3, 98), …, (50, 51). Note that each pair has the sum of 101, and there are 50 pairs, so the sum is equal to 101(50) = 5,050.

In general, we can sum up the numbers 1 to n in a similar pairing process. When n is even, we can pair $(1, n)$, $(2, n - 1)$, $(3, n - 2)$, and so on to make $n/2$ pairs that sum to $n + 1$. When n is odd, we can pair $(0, n)$, $(1, n - 1)$, $(2, n - 2)$ to make $(n + 1)/2$ pairs sum to n. Either way, the formula turns out to be the same $n(n + 1)/2$.

Practice Problems

Sum the whole numbers up to 100.

Sum the whole numbers up to 1,016.

Proof

The formula follows from adding the formulas for the sum of the even numbers and the formula for the sum of the odd numbers, which were proven in the last two sections.

Alternately, the formula can be proven inductively. The formula holds for base cases $n = 1$, sum of $1(2)/2 = 1$, and $n = 2$, sum of $2(3)/2 = 3$. Assuming it is true for the first k numbers, we wish to prove the formula is true for the next term by evaluating $1 + 2 + 3 + \ldots + k + (k + 1)$.

By the induction hypothesis, we know the sum up to k is equal to the value $k(k + 1)$. So we have the following.

$$1 + 2 + 3 + \ldots + k + (k + 1) = k(k + 1)/2 + (k + 1)$$

$$= [k(k + 1) + 2(k + 1)]/2 = (k^2 + 3k + 2)/2 = (k + 1)(k + 2)/2$$

The last step is factoring $k^2 + 3k + 2$, so indeed the formula is verified for adding the next number.

Solutions to Practice Problems

Sum the whole numbers up to 100. By the formula, the sum is $100(101)/2 = 5{,}050$.

Sum the even numbers up to 1,016. By the formula, the sum is $1016(1017)/2$. We can use the trick to multiply numbers near 1,000. The differences are 16 and 17. So the first four digits are found by adding the difference of one number to the other number. That is, we can either do $1016 + 17 = 1033$ or $1017 + 16 = 1033$. Then the final three digits are the product of the differences. This is $16(17) = 16^2 + 17 = 256 + 16 = 272$. So we have 1,033,272, which we divide in half to get 516,636.

Sum Fibonacci Style Sequences

Ask a friend for two numbers between 1 and 9. Let's say the numbers are 1 and 3.

You write out the following list of numbers: 1, 3, 4, 7, 11, 18, 29, 47, 76, 123, 199, 322.

Then your friend picks a number on the list, say 123. You then say the sum of all the numbers in the list up to 123 must be 319. There's no way you could have added up all the numbers that quickly in your head. So how did you do it?

The trick is the numbers in the list follow a pattern. After the first two numbers, every number thereafter is equal to the sum of the previous two numbers. When the list starts with the two numbers 1, 1, the list is then 2, 3, 5, 8, 13, …, which is known as the Fibonacci sequence.

There is a special property to lists generated this way. Specifically, the sum of all the numbers in the list up to a certain point is always equal to the number two spots later minus the second number in the list. In algebra terms:

$$F_1 + F_2 + \ldots + F_n = F_{n+2} - F_2$$

For example, in the list 1, 3, 4, 7, 11, 18, 29, 47, 76, 123, 199, 322, the sum up to any point is always equal to the number two spots later minus the second number in the list of 3. So the sum up to 123 is equal to the number two spots later, 322, minus the second number 3. Similarly, the sum up to 29 will be the number two spots later, 76, minus the second number 3. And you can verify the sum up to 29 is in fact 73.

You can use this trick for a list starting out with any two numbers. You ask for two numbers, and then you generate the rest of the numbers so each new term is the sum of the previous two. Then you can find the sum of all the items up to any point! You just look or calculate two terms ahead and subtract the second item on the list.

Be warned the numbers in the list grow quickly, so you might want to start with some relatively small numbers.

Practice Problems

Sum the numbers in the list 1, 2, 3, 5, 8, 13, 21

Sum the numbers in the list 4, 7, 11, 18, 29, 47

Proof

The formula can be proven inductively. The formula holds for base case $n = 1$ by the definition that $F_1 + F_2 = F_3$, which means $F_1 = F_3 - F_2$. Assuming it is true for the first k numbers, we wish to prove it is true for the next number of $k + 1$.

That is, we want to evaluate the sum $F_1 + F_2 + \ldots + F_k + F_{k+1}$.

By the induction hypothesis, we know the sum up to k is equal to the formula $F_{k+2} - F_2$. So we have the following.

$$F_1 + F_2 + \ldots + F_k + F_{k+1} = F_{k+2} - F_2 + F_{k+1} = F_{k+3} - F_2$$

The last step follows from $F_{k+1} + F_{k+2} = F_{k+3}$. So the formula is verified for adding the next number.

Solutions to Practice Problems

Sum the numbers in the list 1, 2, 3, 5, 8, 13, 21. The next number in the sequence is 13 + 21 = 34 and the one after that is 34 + 21 = 55. Subtracting that term by the second item in the list, 2, gives the sum of 53.

Sum the numbers in the list 4, 7, 11, 18, 29, 47. The next number in the sequence is 29 + 47 = 76 and the one after that is 76 + 47 = 123. Subtracting that term by the second item in the list, 7, gives the sum of 116.

Create A 3x3 Magic Square

A magic square is an arrangement of numbers where every row, column, and diagonal sums to the same number. The classic magic square is a 3x3 grid of the numbers from 1 to 9, where every row, column, and diagonal sums to 15. It is possible to memorize the magic square. But there is a simple trick that will help you generate the magic square if you forget.

We start out with a blank 3x3 grid.

\- - -

\- - -

\- - -

The first step is to place the number 1 in the middle of the top row.

\- 1 -

\- - -

\- - -

Now there is a simple rule. The next number should be placed up and to the right, wrapping around if necessary. If that spot already has a number, then the next number should be written vertically down, again wrapping around if necessary.

So where does the number 2 go? It should be up and to the right of the number 1. Since we cannot move any further up, we wrap around to the bottom row and then move one column to the right. So here is where the number 2 goes, in the bottom right hand corner.

\- 1 -

\- - -

\- - 2

The number 3 is placed up and to the right. We go up to the middle row. But we cannot go further to the right, so we have to wrap around to the first column.

- 1 -

3 - -

- - 2

We can continue the rule. From the 3, we need to go up and to the right. However, that spot already has the number 1. So we need to go down to write the number 4.

- 1 -

3 - -

4 - 2

Now we go up and to the right to write the number 5, so the number 5 is written in the middle column of the middle row.

- 1 -

3 5 -

4 - 2

Then we go up and to the right to write the number 6, so the number 6 is written in the third column of the top row.

- 1 6

3 5 -

4 - 2

From the number 6, going up and to the right, wrapping around vertically and horizontally, means checking the lower left spot.

But the number 4 is already in that spot, so we have to go down from the number 6 to write the number 7 in the third column of the middle row.

- 1 6

3 5 7

4 - 2

Then we go up and to the right to write the number 8 in the first column of the top row.

8 1 6

3 5 7

4 - 2

The final spot, which is also up and to the right, is where the final 9 is, in the middle column of the bottom row.

8 1 6

3 5 7

4 9 2

If you want to mix up the presentation, or avoid drawing attending to the trick, you can stop after writing the numbers 1, 2, and 3 with the rule. Then you should remember the 5 goes in the middle of the middle row, so the magic square looks like this.

- 1 -

3 5 -

- - 2

At this point, remember that each row, column, and diagonal has to sum to 15. This will force the values for the remaining squares. That is, in order to make the middle row sum to 15, the right spot has to be 7. And in order to make the middle column sum to 15, the bottom spot has to be 9.

That is, we can start filling in the square as follows.

- 1 -

3 5 7

- 9 2

From here the top row right number has to be 6, and the bottom row left number has to be 4.

- 1 6

3 5 7

4 9 2

The final spot has to be the 8.

8 1 6

3 5 7

4 9 2

After doing this a couple times you may notice other patterns that will help you remember. For example, the there is a diagonal of 4-5-6 and the middle row is 3-5-7.

- - 6

3 5 7

4 - -

The remaining spots are again forced because the sum has to be 15.

There are several ways to generate a 3x3 magic square; practice with the method that is easiest for you to remember.

Create A 4x4 Magic Square From Your Birthday

A 4x4 magic square has every row, column, and diagonal sum to the same number. The difference is that every row has 4 numbers. A neat trick is you can generate a magic square from a birthday. For example, consider the date 22 December 1887, which was the birthday of the mathematician Srinivasa Ramanujan. We will write this birthday as four two-digit numbers 22-12-18-87 in the format DD-MM-CC-YY being the two-digit values for date, month, century, and year.

The 4x4 magic square has the first row being these numbers.

22 12 18 87

So we want every row, column, and diagonal to sum to 139. There will also be other patterns like many 2x2 blocks will have the same sum, and the sum of the diagonals will be the same as well.

To fill out the rest of the square, you can use the following template.

DD	MM	CC	YY
YY + 1	CC – 1	MM – 3	DD + 3
MM – 2	DD + 2	YY + 2	CC – 2
CC + 1	YY – 1	DD + 1	MM – 1

There are patterns that can help you remember this formula. Every row is based on the day, month, century, and year values in a different order from the first row. Similarly, every column is based on the day, month, century, and year values in a different order in each column.

Furthermore, when we adjust one value up, we pair that by adjusting another value down. For example, in the second row the value YY + 1 is paired with CC – 1, and the value MM – 3 is paired with DD + 3. The adjustments up have to be offset by the adjustments down.

So let's use the template to complete the magic square for Ramanujan's birthday.

22 12 18 87

88 17 09 25

10 24 89 16

19 86 23 11

Every row, column, diagonal sums to 139, and so do the 4 corners. Most of the 2x2 blocks also sum to 139 as well.

Here is another magic square, for the date 04 July 1920, generated using the formula.

04 07 19 20

21 18 04 07

05 06 22 17

20 19 05 06

Here is a final magic square, for the date 14 February 1980, generated using the formula.

14 02 19 80

81 18 -1 17

00 16 82 17

20 79 15 01

You will notice there is a value of -1 in the second row, third column. There is nothing wrong with this mathematically, and the square will still have the same sum of 115 for every row, column, and diagonal.

Convert A Decimal Number To Binary

Most transactions and daily calculations we do involve numbers written in the decimal system. This is a numerical system where each digit of a number represents a power of 10. For example, the last digit of a number is called the "units" digit ($1 = 10^0$), the next digit is called the "tens" digit ($10 = 10^1$), the next is called the "hundreds" digit ($100 = 10^2$), and so on. A number like 1,234 is a shorthand for the following equivalent expression $1,234 = 1(1,000) + 2(100) + 3(10) + 4(1)$.

The decimal system uses the convention of powers of 10. Mathematically, it is possible to use other number bases. Computers typically use powers of 2 and that is called the binary system. The digits of a number can either be 0 or 1, and each placeholder is a power of 2.

For example, what does the number 1101 in the binary system mean in the decimal system? Every digit represents a power of 2, so when we write 1101, we mean $1101 = 1(2^3) + 1(2^2) + 0(2^1) + 1(2^0)$. We can evaluate the sum which will give an equivalent representation in the decimal number system. That is, we can calculate that the number 1101 means $1101 = 1(2^3) + 1(2^2) + 0(2^1) + 1(2^0) = 8 + 4 + 1 = 13$ in the more familiar decimal system. So 1101 in binary is 13 in decimal.

Just as we converted 1101 from binary to decimal, we can convert any binary number to its equivalent decimal representation. But how can we go in the other direction? That is, how can we convert a number written in the decimal system to its equivalent binary system representation?

There is a shortcut trick to speed up the calculation. What we do is repeatedly divide a number in half, ignoring any remainders, and write each result to the left of the previous number. We continue until we reach the number 1. As that point, we replace each even number with a 0, and replace each odd number with a 1. The result, read from left to right, is the number written in binary.

Let's do an example of converting the decimal number 13 into binary. When we divide 13 in half, we have 6 with a remainder of 1. We ignore the remainder and write the number 6 to the left of 13.

6, 13

Then we repeat the process and divide 6 in half. This results in 3, so we write 3 to the left of 6.

3, 6, 13

Finally we divide 3 in half, which is 1 with a remainder of 1. We ignore the remainder and write 1.

1, 3, 6, 13

Now we have reached 1 so we can stop. Finally we replace each even number with a 0 and each odd number with a 1. So let's write this out.

1, 3, 6, 13 → 1 1 0 1

This means the number 1101 is the number 13 converted to binary. And this matches what we already knew when we converted 1101 from binary to its decimal equivalent.

(Because $1101 = 1(2^3) + 1(2^2) + 0(2^1) + 1(2^0) = 8 + 4 + 1 = 13$.)

Let's do another example of converting the decimal number 2015 into its binary equivalent.

Repeatedly halving 2015 and ignoring the remainder results in 1007, 503, 251, 125, 62, 31, 15, 7, 3, 1. We write each number to the left to get the following sequence.

1, 3, 7, 15, 31, 62, 125, 251, 503, 1007, 2015

Now we replace each even number with a 0 and each odd number with a 1.

1, 3, 7, 15, 31, 62, 125, 251, 503, 1007, 2015 → 1 1 1 1 1 0 1 1 1 1 1

So 11111011111 is the number 2015 written in binary, and you can check this is a correct representation of the number 2015 in binary by going in reverse and adding up the powers of 2.

It is relatively easy to halve numbers mentally, so with practice you can fairly easily convert decimal numbers to binary in your head.

Practice Problems

Convert the following decimal numbers into binary.

12

95

132

2,840

39,951

Proof

Suppose N is a number in the decimal system. We want to express N in the binary system as the sum of powers of 2. That is, we want to find out $N = n_0 2^0 + n_1 2^1 + \ldots + n_{k-1} 2^{k-1} + n_k 2^k = n_0 + n_1 2^1 + \ldots + n_{k-1} 2^{k-1} + n_k 2^k$.

How can we obtain the coefficients attached to the powers of 2?

Let us divide N repeatedly by 2 a total of k times.

$N = n_0 + [n_1 2^1 + \ldots + n_{k-1} 2^{k-1} + n_k 2^k]$

$N/2 = n_1 + [n_0 2^{-1} \ldots + n_{k-1} 2^{k-2} + n_k 2^{k-1}]$

$N/2^2 = n_2 + [n_0 2^{-2} \ldots + n_{k-1} 2^{k-3} + n_k 2^{k-2}]$

…

$N/2^k = n_k + [n_0 2^{1-k} \ldots + n_{k-1} 2^{-1}]$

Now let's consider whether the whole parts for each line, rounded down, are odd or even. We claim the answer depends only on the first term n_j in the summation. Why? First we will delete the fractional "remainder" parts of each line, because we are only interested in whether the whole part is even or odd. That is, we can delete the terms multiplied by 2 raised to a negative power. Second, we can delete any terms that are even. The reason is even + even = even, and odd + even = odd. So if we delete the even terms, we do not change whether the whole part is even or odd.

Therefore, to find out whether the whole part of a given line is even or odd, we can delete the entire bracketed term of each line and only consider the first term n_j. The first number n_j will either be 0—making the number $N/2^j$ even—or it will be 1—making the number $N/2^j$ odd.

So let's put this all together. Repeatedly dividing by 2, and then writing the next term to the left results in the terms $[N/2^k, N/2^{k-1}, \dots N/2^2, N/2, N]$. When we ignore the fractional parts, and then only consider whether the terms are even or odd, that only depends on the evenness or oddness of following terms $[n_k, n_{k-1}, \dots, n_2, n_1, n_0]$. Therefore, by repeatedly dividing by 2, and replacing even numbers with 0 and odd numbers with 1, we exactly obtain the coefficients for the binary representation of N.

Solutions to Practice Problems

12. Repeatedly dividing by 2, ignoring the remainders, and writing each term to the left, gives 1, 3, 6, 12. Writing a 1 for each odd number and 0 for each even number, we get 1, 1, 0, 0. So this is 1100 in binary.

95. Repeatedly dividing by 2, ignoring the remainders, and writing each term to the left, gives 1, 2, 5, 11, 23, 47, 95. Writing a 1 for each odd number and 0 for each even number, we get 1, 0, 1, 1, 1, 1, 1. So this is 1011111 in binary.

132. Repeatedly dividing by 2, ignoring the remainders, and writing each term to the left, gives 1, 2, 4, 8, 16, 33, 66, 132. Writing a 1 for each odd number and 0 for each even number, we get 1, 0, 0, 0, 0, 1, 0, 0. So this is 10000100 in binary.

2,840. Repeatedly dividing by 2, ignoring the remainders, and writing each term to the left, gives 1, 2, 5, 11, 22, 44, 88, 177, 355, 710, 1420, 2480. Writing a 1 for each odd number and 0 for each even number, we get the binary number 101100011000.

39,951. Repeatedly dividing by 2, ignoring the remainders in each step, and writing each term to the left gives 1, 2, 4, 9, 19, 39, 78, 156, 312, 624, 1248, 2496, 4993, 9987, 19975, 39951. Writing a 1 for each odd number and 0 for each even number, we get the binary number 1001110000001111.

The Egyptian Method / Russian Peasant Multiplication

What is 13 x 24? There are a number of ways to solve this problem. The Egyptian Method, also known as Russian Peasant Multiplication, is a fascinating method that only uses the process of halving, doubling, and adding to solve multiplication problems.

The first step is to repeatedly halve the first number, ignoring the remainders, until we get to 1. This is the same process used in the section about converting a number to binary.

So to do 13 x 24, we will repeatedly halve 13, ignoring the remainders. We get 6, 3, and 1. We write these numbers under the 13, each number in a new row.

13 24

6

3

1

The second step is to repeatedly double the second number, writing each new number in the row below the previous one. So to double 24 we get 48, 96, and 192.

13 24

6 48

3 96

1 192

Now we have a table with two columns. We will need to inspect the values in the first column.

The third step is to cross out any rows where the number in the first column is even. Out of 13, 6, 3, and 1, only the number 6 is even. So we cross out the entire row that starts with the number 6.

13 24

~~6 48~~

3 96

1 192

Finally, we add up the numbers in the second column that have not been crossed out. Thus, we add up 24, 96, and 192. The result is 312. And somewhat remarkably, 312 is the correct answer to the problem of 13 times 24!

To summarize, there are four main steps to multiply numbers X and Y by the Egyptian Method or Russian Peasant Multiplication.

1. Repeatedly halve the number X, ignoring remainders, until you get to the number 1. Write each new number in the same column as X, with each new number starting another row.

2. Correspondingly double the number Y repeatedly, writing each new number in the row below the previous one, and lining up with a corresponding halved value of X.

3. Look at the first column where X has been halved repeatedly. For each even number in the column, cross out the numbers in that entire row.

4. Add up the values in the second column with the doubled Y values that have not been crossed out. This is the result of X times Y.

One issue is when you multiply two numbers, you can choose which number to make X. Personally I choose X to be the smaller number because that will mean fewer steps of halving and ultimately fewer numbers to double and add. For example, when we did 13x24 we could have made X to be 13 or 24, and we chose the smaller number $X = 13$.

A second thing to remember is sometimes the first row will have an even number. In that case, you must cross out that row as well. For example, let's do 14x24. When we repeatedly halve 14, we get the values 7, 3, and

1. And repeatedly doubling 24 gets the values of 48, 96, and 192. So our table is the following.

14 24

7 48

3 96

1 192

In the first column, only the value 14 is an even number. That means we should cross out the entire first row.

~~14 24~~

6 48

3 96

1 192

Thus we need to add up 48, 96, and 192, which leads to the result of 336. This is indeed the answer to 14 times 24.

The Egyptian Method / Russian Peasant Multiplication is a pretty cool trick because you can multiply any two numbers with the operations of halving, doubling, and adding.

Practice Problems

Solve the following problems using the Egyptian Method / Russian Peasant Multiplication.

12 x 16

102 x 126

Proof

So why does it work? The short answer is the procedure works because we are essentially doing multiplication in binary. The process of dividing

X in half repeatedly, ignoring remainders, is the same procedure described in the section about converting a number from the decimal system to the binary system. The only difference is we are writing the numbers below each other vertically. Also, instead of writing 1s for odd and 0s for even, we mark the even number 0s by crossing out the rows.

The process of repeatedly doubling Y is about attaching the appropriate power of 2 for the binary number. Adding up the doubled values has the effect of finding the decimal equivalent for the product.

Let's do an example to give an intuition for the proof. To do 13x24, we first are repeatedly halving 13 to get 6, 3, and 1. The binary number is obtained by considering the numbers in the order (1, 3, 6, 13), putting 0s for even numbers, and 1s for odd numbers. So converting 13 to a binary number is $1101 = 1(2^3) + 1(2^2) + 1(2^1) + 1(2^0)$.

Consider multiplying this number by 24. We can multiply the right-hand side expression to get $13\text{x}24 = [1(2^3) + 1(2^2) + 1(2^1) + 1(2^0)]\ (24)$. When we distribute the 24, we have $13\text{x}24 = 24(2^3) + 24(2^2) + 0(2^1) + 24(2^0)$.

In other words, we can do 13x24 by adding 24 with no doubling (2^0), plus 24 when doubled two times (2^2), plus 24 doubled three times (2^3). And we cross out the row where 24 is doubled only once, because the binary value for 2^1 was a 0.

The Egyptian Method / Russian Peasant Multiplication is a means to do this procedure in a table form.

Solutions to Practice Problems

12 x 16. We will set up a table where the first row has the numbers 12 and 16. Then we will repeatedly divide 12 in half, ignoring remainders.

12 16

6

3

1

Then we repeatedly double 16, writing each new number in a row below the previous one, lining up with the corresponding halved values of 12.

12 16

6 32

3 64

1 128

At this point we inspect the first row for any even number. We cross out the numbers in the entire row when there is an even number in the first column. The even numbers are 12 and 6, so we cross out the first and second rows.

~~12 16~~

~~6 32~~

3 64

1 128

Finally, we add up the values not crossed out, 64 and 128, to get 192.

102 x 126. We repeatedly divide 102 in half, ignoring remainders.

102 126

61

30

15

7

3

1

Then we repeatedly double 126.

102 126

51 252

25 504

12 1,008

6 2,016

3 4,032

1 8,064

Then we cross out any rows that begin with an even number.

~~102 126~~

51 252

25 504

~~12 1,008~~

~~6 2,016~~

3 4,032

1 8,064

Finally we add up the numbers in the right hand column that are not crossed out. Adding up 252, 504, 4,032, and 8,064 gives the result 12,852.

Extract Cube Roots

Here's a fun trick you can try. Ask a friend to pick a whole number between 1 and 100 and keep the number a secret. On a calculator, the friend cubes the number, which means raising to the third power, or multiplying the number by itself twice. Ask for the number. Let's say it is 157,464. You instantly do the cube root in your head and say the original number was 54. How is this possible?

The trick is that the cubes of two-digit numbers follow a specific pattern. But in order to pull off the trick, you first need to memorize the cubes of numbers from 1 to 9. Here are the cubes.

$1^3 = 1$

$2^3 = 8$

$3^3 = 27$

$4^3 = 64$

$5^3 = 125$

$6^3 = 216$

$7^3 = 343$

$8^3 = 512$

$9^3 = 729$

The other part you need to memorize is the units digit for each of these cubes. Note that for the cubes of 1, 4, 5, 6, and 9, the units digit is the same number. Also the cubes for the pairs of numbers 2, 8 and 3, 7 have last digits that swap (The cube of 2 ends in 8, the cube of 8 ends in 2; the cube of 3 ends in 7, and the cube of 7 ends in 3).

Now you're ready to solve for the cube roots when someone cubes a two-digit number.

The cube of any two-digit number is always between 4 to 6 digits. In other words, the cube is always between 1,000 and 999,999. For this trick, it will be sufficient to focus the digits above the thousands (left of the comma, or excluding the last three digits) and the digit in the units spot (the last digit). So let's write the cube of a two-digit number number as $X...Y$ for the two relevant digits of the number above the thousands spot X and in the units spot Y.

Here is the procedure. We look at the part above the thousands, X. The number X will be between two of the cubes for single digit numbers we have memorized. The lower number becomes the first digit of our answer. (Or X may be larger than 9^3, in which case $X = 9$).

We then look at the last digit of the number, Y. We look for the single-digit cube whose last digit matches Y. That becomes the final digit in our answer.

Here is the procedure using the example 157,464. The part above the thousands, X, is equal to 157. So we think about the cubes we have memorized. The number 157 is between $5^3 = 125$ and $6^3 = 216$. We take the lower number, 5, and that's the first digit in our answer.

The last digit of the number, Y, is equal to 4. The cube whose last digit is 4 is 4 (because $4^3 = 64$). That becomes the final digit in our answer. Therefore, we conclude the answer is 54. The surprising part is we can extract the cube root without having to do any difficult calculations. We simply rely on the memorized values and we can determine the original number.

Let's do another example of 658,503. The part above the thousands, X, is equal to 658. This number is between the cubes $8^3 = 512$ and $9^3 = 729$. We take the lower cube, 8, and that's the first digit in our answer.

The last digit of the number, Y, is equal to 3. The cube whose last digit is 3 is equal to 7 (because $7^3 = 343$). That becomes the final digit in our answer. Therefore, we conclude the answer is 87.

Extracting cube roots is an impressive trick because it is not obvious how you can quickly solve a very difficult problem. The trick is all about memorizing the value of the single digit cubes and then focusing on specific parts of the cube. We'll generalize this trick in the next sections.

Practice Problems

Extract the cube roots of the following numbers.

2,744

15,625

29,791

110,592

389,017

Proof

Why does this trick work? The procedure is to figure out each of two digits in a number that is cubed (raised to the third power). Denote the number $AB = 10A + B$. The goal is to determine A and B upon hearing the value for $AB^3 = (10A + B)^3$.

The first step is to figure out A. This is done by estimation. The idea is to consider the multiples of ten, the numbers 10, 20, 30, …, 90, raised to the third power. That is, we want to memorize 10^3, 20^3, 30^3, …, 90^3 as estimation ranges.

How does this help? Note that any number between 10 and 19 raised to power of 3 will be between 10^3 and 20^3, any number between 20 and 29 raised to power of 3 will be between 20^3 and 30^3, and so on. So if we can memorize the multiples of 10 raised to the power of 3, then we can readily figure out the first digit A of the answer. So how do we do that?

The trick is we memorize the numbers 1, 2, 3, …, 9 each raised to the power of 3. Then we can deduce the list of 10^3, 20^3, 30^3, …, 90^3 as the same numbers with an additional 3 zeros. But rather than memorizing all these numbers, we employ another trick. When someone tells us the number raised to the power of 3, we ignore the last 3 digits of the number. This allows us to determine which two multiples of 10 the number is between, and that in turn allows us to deduce the first digit of the answer A by the process of estimation.

Then we need to figure out the second digit B. This part is easier. The last digit of the number must end in the same digit as B^3. This is intuitive, but we can prove it algebraically as well. Note that $AB^3 = (10A + B)^3$, which is equal to $1000A^3 + 300A^2B + 300AB^2 + B^3$. Every term except B^3 is multiplied by a power of 10, so only the term B^3 affects the units digit of the cube of *AB*.

Then we note a pattern to the last digit of the cubes. The cubes of 1, 4, 5, 6, and 9 end in the same digit, whereas the cube of 2 ends in 8, the cube of 3 ends in 7, the cube of 7 ends in 3, and the cube of 8 ends in 2.

Solutions to Practice Problems

2,744. Deleting the last 3 digits results in the number 2. This is between $1^3 = 1$ and $2^3 = 8$, so the first digit of the cube root is the lower number 1. Then the last digit of 2,744 is 4. The last digit is 4 for the cube of 4, and hence the second digit is 4. The cube root is 14.

15,625. Deleting the last 3 digits results in the number 15. This is between $2^3 = 8$ and $3^3 = 27$, so the first digit of the cube root is the lower number 2. Then the last digit of 15,625 is 5. The last digit is 5 for the cube of 5, and hence the second digit is 5. The cube root is 25.

29,791. Deleting the last 3 digits results in the number 29. This is between $3^3 = 27$ and $4^3 = 64$, so the first digit of the cube root is the lower number 3. Then the last digit of 29,791 is 1. The last digit is 1 for the cube of 1, and hence the second digit is 1. The cube root is 31.

110,592. Deleting the last 3 digits results in the number 110. This is between $4^3 = 64$ and $5^3 = 125$, so the first digit of the cube root is the lower number 4. Then the last digit of 110,592 is 2. The last digit is 2 for the cube of 8, and hence the second digit is 8. The cube root is 48.

389,017. Deleting the last 3 digits results in the number 389. This is between $7^3 = 343$ and $8^3 = 512$, so the first digit of the cube root is the lower number 7. Then the last digit of 389,017 is 7. The last digit is 7 for the cube of 3, and hence the second digit is 3. The cube root is 73.

Extract Fifth Roots

We can modify the cube root trick to extract fifth roots. Ask a friend to pick a whole number between 10 and 100 and keep the number a secret. On a calculator, the friend raises the number to the fifth power. Ask for the number. Let's say it is 69,343,957. You instantly do the fifth root in your head and say the original number was 37. How is this possible?

The trick is that the fifth power of two-digit numbers can be solved from two pieces of information. But in order to pull off the trick, you first need to memorize the fifth powers of the numbers from 1 to 9.

$1^5 = 1$

$2^5 = 32$

$3^5 = 243$

$4^5 = 1,024$

$5^5 = 3,125$

$6^5 = 7,776$

$7^5 = 16,807$

$8^5 = 32,768$

$9^5 = 59,049$

The other part you need to memorize is the last digit of each of these fifth powers. This is easy: the last digit is the same as the number you are raising to the fifth power.

Now you're ready to solve for the fifth root when someone raises a two-digit number to the fifth power.

The fifth power of any two-digit number is between 6 to 10 digits. In other words, the cube is between 100,000 and 9,999,999. For this trick, it will be useful to consider the digits above the hundred thousands (ignore the last 5 digits of the number) and the digit in the units spot (the last digit). So let's write the fifth power of a two-digit number number as $X...Y$ for the two relevant digits of the number.

Here is the procedure. We look at the part above the hundred thousands, X. The number X will be between two of the fifth powers for single digit numbers we have memorized. Whichever number corresponds to the lower fifth power is the first digit of the answer. (Or X may be larger than 9^5, in which case $X = 9$).

We then look at the last digit of the number, Y. That will be the second digit of the answer.

Here is the procedure using the example 69,343,957. The part above the hundred thousands, X, is equal to 693. This number is between the two fifth powers of $3^5 = 243$ and $4^5 = 1{,}024$. We take the lower fifth power, 3, and that's the first digit in our answer.

The last digit of the number, Y, is equal to 7. This is the second digit of the answer. Therefore, we conclude the fifth root is 37.

Let's do another example of 459,165,024. The part above the hundred thousands, X, is equal to 4,591. This number is between the two fifth powers of $5^5 = 3{,}125$ and $6^5 = 7{,}776$. We take the lower fifth power, 5, and that's the first digit in our answer.

The last digit of the number, Y, is equal to 4. That becomes the final digit in our answer. Therefore, we conclude the fifth root is 54.

Practice Problems

Extract the fifth roots of the following numbers.

537,824

28,629,151

550,731,776

844,596,301

1,934,917,632

Proof

This trick works for a similar reason to the trick for extracting cube roots, but we'll explain the steps in detail again for completeness. The procedure is to figure out each of two digits in a number that has been raised to the fifth power. Denote the number $AB = 10A + B$, so the goal is to find A and B when told the value $AB^5 = (10A + B)^5$.

The first step is to figure out A. This is done by estimation. The idea is to consider the multiples of ten, the numbers 10, 20, 30, …, 90, raised to the fifth power. So we memorize 10^5, 20^5, 30^5, …, 90^5 to create ranges. Now any number between 10 and 19 raised to power of 5 will be between 10^5 and 20^5, any number between 20 and 29 raised to power of 5 will be between 20^5 and 30^5, and so on. So if we could memorize the multiples of 10 raised to the power of 5, then we can readily figure out the first digit A of the answer. So how do we do that?

The trick is we memorize the numbers 1, 2, 3, …, 9 each raised to the power of 5. Then we can deduce the list of 10^5, 20^5, 30^5, …, 90^5 as the same numbers with an additional 5 zeros. But rather than memorizing all these numbers, we can equivalently ignore the last 5 digits when someone tells us a number. This allows us to determine which two multiples of 10 the number is between, and that allows us to deduce the first digit of the answer A.

Then we need to figure out the second digit B. This part is easier. The last digit of the number must end in the same digit as B^5. This is from the fact that $AB^5 = (10A + B)^5$. If we expanded this out, every term except B^5 would be multiplied by a power of 10, so only the term B^5 affects the units digit of the fifth power of AB. There is a nice pattern for fifth powers: the last digit is the same as the original number.

Solutions to Practice Problems

537,824. Deleting the last 5 digits results in the number 5. This is between $1^5 = 1$ and $2^5 = 32$, so the first digit of the fifth root is the lower number 1. Then the last digit of the number is 4, which means the original number had a last digit of 4 as well. The fifth root is 14.

28,629,151. Deleting the last 5 digits results in the number 286. This is between $3^5 = 243$ and $4^5 = 1,024$, so the first digit of the fifth root is the lower number 3. Then the last digit of the number is 1, which means the original number had a last digit of 1 as well. The fifth root is 31.

550,731,776. Deleting the last 5 digits results in the number 5,507. This is between $5^5 = 3{,}125$ and $6^5 = 7{,}776$, so the first digit of the fifth root is the lower number 5. Then the last digit of the number is 6, which means the original number had a last digit of 6 as well. The fifth root is 56.

844,596,301. Deleting the last 5 digits results in the number 8,445. This is between $6^5 = 7{,}776$ and $7^5 = 16{,}807$, so the first digit of the fifth root is the lower number 6. Then the last digit of the number is 1, which means the original number had a last digit of 1 as well. The fifth root is 61.

1,934,917,632. Deleting the last 5 digits results in the number 19,349. This is between $7^5 = 16{,}807$ and $8^5 = 32{,}768$, so the first digit of the fifth root is the lower number 7. Then the last digit of the number is 2, which means the original number had a last digit of 2 as well. The fifth root is 72.

Extract Odd-Powered Roots

Are the tricks for taking cube roots and fifth roots special? Actually, we can generalize the procedure to extract any odd power, so we can extract the roots for seventh roots, ninth roots, and so on, for any two-digit number.

Odd powers are numbers of the form $2k + 1$. Here is the general procedure to extract odd roots when someone raises a two-digit number to the power of $2k + 1$.

1. Memorize the numbers 1 to 9 raised to the power of $2k + 1$. You have to memorize different values for each power that you want to perform the trick.

2. Ask someone to raise a secret number to the power of $2k + 1$. You only pay attention to two parts of the digits. You ignore or delete the last $2k + 1$ digits to make the number X. Then you consider last digit as Y.

3. The number X will be between two of the powers for numbers you have memorized. Pick the lower number, and that is the first digit of your answer. (Or X may be larger than 9^{2k+1}, in which case $X = 9$).

4a. If the last digit Y is 1, 4, 5, 6, or 9, then that is the second digit of your answer.

4b. If Y is 2, 3, 7, or 8, then it depends on the power $2k + 1$. When the power is 5, 9, 13, etc. (one more than a multiple of 4), then the second digit is the same as Y. When the power is 3, 7, 11, etc. (one less than multiple of 7), then it's the same procedure as the one for cube roots. The pairs 2, 8 and 3,7 will swap. That is, when Y is 2, the last digit is 8; when Y is 8, the last digit is 2; when Y is 3, the last digit is 7; and when Y is 7, the last digit is 3.

The amazing part is this procedure works for any odd powers, and it is based on the same method as extracting cube roots and fifth roots.

Let's do an example for extracting the 7th root of a two-digit number raised to the power of 7.

We first need to memorize these powers.

$1^7 = 1$

$2^7 = 128$

$3^7 = 2,187$

$4^7 = 16,384$

$5^7 = 78,125$

$6^7 = 279,936$

$7^7 = 823,543$

$8^7 = 2,097,152$

$9^7 = 4,782,969$

With the 7th root, the power of 7 is one less than a multiple of 4, we have that the numbers 1, 4, 5, 6, and 9 end in the same last digit, but the pairs 2, 8 and 3, 7 swap last digits.

Now you're ready to solve for the seventh root when someone raises a two-digit number to the seventh power.

Let's say someone tells you 10,460,353,203 is the result of raising a two-digit number to the seventh power. We ignore the last 7 digits to find, *X*, which is equal to 1,046. This number is between the two seventh powers of $2^7 = 128$ and $3^7 = 2,187$. We therefore take the lower seventh power, 2, and that's the first digit in our answer.

The last digit of the number, *Y*, is equal to 3. In the seventh powers, the last digit of 3 corresponds to a swap with the number 7. So the last digit is 7. Therefore, we conclude the answer is 27.

Practice Problems

Extract the 7th root of 3,404,825,447

Extract the 9th root of 5,159,780,352

Proof

We generalize the proof for the ones given in the cube root and fifth root sections for the power $2k + 1$. Denote the number $AB = 10A + B$, so the goal is to find A and B from the number $AB^{2k+1} = (10A + B)^{2k+1}$.

The first step is to figure out A. This is done by estimation. The idea is to consider the multiples of ten, the numbers 10, 20, 30, …, 90, raised to the power of $2k + 1$. So we memorize 10^{2k+1}, 20^{2k+1}, 30^{2k+1}, …, 90^{2k+1} as ranges. Now any number between 10 and 19 raised to $2k + 1$ will be between 10^{2k+1} and 20^{2k+1}, any number between 20 and 29 raised to $2k + 1$ will be between 20^{2k+1} and 30^{2k+1}, and so on. So if we could memorize the multiples of 10 raised to the power of $2k + 1$, then we can readily figure out the first digit A of the answer. So how do we do that?

The trick is we memorize the numbers 1, 2, 3, …, 9 each raised to the power of $2k + 1$. Then we can deduce 10^{2k+1}, 20^{2k+1}, 30^{2k+1}, …, 90^{2k+1} as the same numbers with an additional $2k + 1$ zeros. But rather than memorizing all these numbers, we employ another trick. When someone tells us the number raised to the power of $2k + 1$, we can ignore the last $2k + 1$ digits of the number. This allows us to determine which two multiples of 10 the number is between, and that allows us to deduce the first digit of the answer A.

Then we need to figure out the second digit B. When the power is 5, 9, 13, …, or one more than a multiple of 4, the second digit B is exactly the same as the last digit of the number. When the power is 3, 7, 11, …, or one less than a multiple of 4, the last digits of 1, 4, 5, 6, 9 are the same as B, and the last digits of 2, 8 and 3, 7 are pairs that swap.

Here is why this happens. When we raise $AB = 10A + B$ to the power of $2k + 1$, that is the same as expanding $(10A + B)^{2k+1}$ by the binomial theorem. The terms involving A will all be multiplied by a factor of 10, so they will all be shifted over to the tens spot or further. So last digit of this number will be the same as the last digit of B^{2k+1}, which we have memorized.

There is a pattern in the last digit of B^{2k+1}. The odd-powers of 1, 4, 5, 6, and 9 all end in the same digit. The odd-powers of 2, 3, 7, and 8 are the same for the fifth powers, but they swap as pairs 2, 8 and 3, 7 for cubes.

So why does the pattern continue, where the powers 7, 11, …, follow the same pattern as cubes, and the powers 9, 12, …, follow the same pattern as fifth powers?

First, consider the 4th powers of the numbers 1 to 9 and then focus on only the last digit (that is, take the result modulo 10).

$1^4 = 1 = 1 \pmod{10}$

$2^4 = 16 = 6 \pmod{10}$

$3^4 = 81 = 1 \pmod{10}$

$4^4 = 256 = 6 \pmod{10}$

$5^4 = 625 = 5 \pmod{10}$

$6^4 = 1{,}296 = 6 \pmod{10}$

$7^4 = 2{,}401 = 1 \pmod{10}$

$8^4 = 4{,}096 = 6 \pmod{10}$

$9^4 = 6{,}561 = 1 \pmod{10}$

We can find the last digit of a number raised to the power of 7 by multiplying its cube by the 4th power and then considering the result modulo 10. Similarly, we can find the last digit of the same number raised to the power of 11, 15, and so on, by successively multiplying the same number raised to the 4th power and then considering the result modulo 10.

Similarly, we can find the last digit of a number raised to the powers of 9, 11, 15, and so on by successively multiplying the fifth power by the same number raised to the 4th power and then considering the result modulo 10.

For example, let's consider the number 2. The cube of 2 is 8, and the 4th power of 2 is 6 (mod 10). Therefore, the last digit of the 7th power is equal to the product of 2 and 6 taken modulo 10, so that gives us the result $2(6) = 2 \pmod{10}$. So the last digit of 2^7 is a 2.

We can similarly calculate the odd powers for each number between 1 and 9. Using the values for cubes, fifth powers, and fourth powers, we can calculate the following values modulo 10.

$1^3 = 1^5 = 1^7 = \ldots = 1 \pmod{10}$

$4^3 = 4^5 = 4^7 = \ldots = 4 \pmod{10}$

$5^3 = 5^5 = 5^7 = \ldots = 5 \pmod{10}$

$6^3 = 6^5 = 6^7 = \ldots = 6 \pmod{10}$

$9^3 = 9^5 = 9^7 = \ldots = 9 \pmod{10}$

$2^5 = 2^9 = 2^{4k+1} = 2 \pmod{10}$ and $2^3 = 2^7 = 2^{4k-1} = 8 \pmod{10}$

$3^5 = 3^9 = 3^{4k+1} = 3 \pmod{10}$ and $3^3 = 3^7 = 3^{4k-1} = 7 \pmod{10}$

$7^5 = 7^9 = 7^{4k+1} = 7 \pmod{10}$ and $7^3 = 7^7 = 7^{4k-1} = 3 \pmod{10}$

$8^5 = 8^9 = 8^{4k+1} = 8 \pmod{10}$ and $8^3 = 8^7 = 8^{4k-1} = 2 \pmod{10}$

Thus, there emerge two patterns: one for raising to the 3rd power, and another for raising to the 5th power, which repeats when adding multiples of 4 to those powers. That is where we get the rule for how to determine the last digit when raising a number to the power of $2k + 1$.

Solutions to Practice Problems

Extract the 7th root of 3,404,825,447. Deleting the last 7 digits results in the number 340. This is between $2^7 = 128$ and $3^7 = 2,187$, so the first digit is the lower number 2. Then the last digit of the number is 3. To extract a 7th root, which is one less than a multiple of 4, we have the last digits for the pair 3 and 7 swap. So the original number had a last digit of 3. The seventh root is 23.

Extract the 9th root of 5,159,780,352. Deleting the last 9 digits results in the number 5. This is between $1^9 = 1$ and $2^9 = 512$, so the first digit is the lower number 1. Then the last digit of the number is 2. To extract a 9th root, which is one more than a multiple of 4, we have the last digit of the original number will be the same. So the original number had a last digit of 2. The ninth root is 12.

Conclusion

I hope you enjoyed learning these math tricks and methods for faster calculation. Some of the tricks like calculating percentages, re-grouping factors, or calculating the day of the week will help on a regular basis as these are common problems. Other tricks like dividing by 91 or multiplying two numbers in the 90s are fun methods that turn very difficult problems into simple ones. And other tricks like dividing by 19 or making a magic square are surprising results that illustrate unexpected patterns and serve as an invitation to learn more math. I hope these tricks can make your life easier, illustrate how math makes hard problems easy, and are entertaining to share to inspire others to learn some fun math as well.

More From Presh Talwalkar

I hope you enjoyed this book. If you have a comment or suggestion, please email me presh@mindyourdecisions.com

The Joy of Game Theory: An Introduction to Strategic Thinking. Game Theory is the study of interactive decision-making, situations where the choice of each person influences the outcome for the group. This book is an innovative approach to game theory that explains strategic games and shows how you can make better decisions by changing the game.

Math Puzzles Volume 1: Classic Riddles in Counting, Geometry, Probability, and Game Theory. This book contains 70 interesting brain-teasers.

Math Puzzles Volume 2: What do an Infinite Tower, a Classic Physics Puzzle, and Coin Flipping Have in Common? This is a follow-up puzzle book with 45 delightful puzzles.

But I only got the soup! This fun book discusses the mathematics of splitting the bill fairly.

40 Paradoxes in Logic, Probability, and Game Theory. Is it ever logically correct to ask "May I disturb you?" How can a football team be ranked 6th or worse in several polls, but end up as 5th overall when the polls are averaged? These are a few of the thought-provoking paradoxes covered in the book.

Multiply By Lines. It is possible to multiply large numbers simply by drawing lines and counting intersections. Some people call it "how the Japanese multiply" or "Chinese stick multiplication." This book is a reference guide for how to do the method and why it works.

About The Author

Presh Talwalkar studied Economics and Mathematics at Stanford University. His site *Mind Your Decisions* has blog posts and original videos about math that have been viewed millions of times.

Printed in Great Britain
by Amazon